PIANOS,
PIANO TUNERS
AND THEIR PROBLEMS

PIANOS, PIANO TUNERS AND THEIR PROBLEMS

GEORGE W. BOOTH

JANUS PUBLISHING COMPANY
London, England

First published in Great Britain 1996
by Janus Publishing Company
Edinburgh House, 19 Nassau Street
London W1N 7RE

Copyright © George W. Booth 1996

British Library Cataloguing-in-Publication Data
A catalogue record of this book is available
from the British Library

ISBN 1 85756 215 1

All rights reserved. No part of this publication
may be reproduced, stored in a retrieval system,
or transmitted in any form or by any means,
electronic, mechanical, photocopying, recording
or otherwise, without the prior permission of the
publisher.

The right of George W. Booth to be identified as
the author of this work has been asserted by him in
accordance with the Copyright Designs and Patents Act 1988.

Cover design Linda Wade

Printed and bound in England
by Antony Rowe
Chippenham, Wiltshire

CONTENTS

Introduction	ix
Foreword	xi
Acknowledgements	xv

1. **CASE HISTORIES** — 1
   - Split wrestplank — 1
   - Broken hammer shanks — 3
   - Blüthner grand 1934 sticking notes — 4
     - Over-enthusiastic use of glue — 4
     - Signs of damp — 6
     - Signs of mice in upright pianos — 7
     - Mouse damage in grand pianos — 7
     - Warped lock rail or key slip — 9
   - Steinway grand 1921, noisy pedals — 10
   - Steinway upright 1894, tuning problem — 11
   - Steinway upright 1937 wear problem — 13
   - Dagmar upright 1978, noisy keys — 14
   - AMYL grand, sticking notes — 15
   - Challen upright 1936, faint wire sounds — 17
   - Loose hammer heads — 18
   - The 'Brasso' man — 19

2. **REPAIRS** — 23
   - Chatter: cause and effect — 24
   - Split bass bridge — 28
   - Water damage — 31
   - 'Dog-leg' key repair — 32
   - Grand leg repair — 34
   - Coated steel strings — 35
   - Synthetic key coverings — 36
   - Refacing/Reshaping of hammer heads — 39
   - Respacing/Adjustment of hammer travel — 42

3. 'D' TYPE SPRING AND LOOP GRAND PIANO
   ACTIONS   45
   Poor quality materials   45
   Warped key slip or lock rail   46
   Repairing broken loops   47
   Pedal adjustment   48
   Loose flange rails (beams)   49
   Alignment of flange rails (beams)   49
   Variations on a theme   51

4. MINIATURE PIANOS   55
   Eavestaff Minipiano 1936 Tuning   55
     Access to the action   58
   Eavestaff Minipiano 1937   61
     Access to the action   61
   Eavestaff Minipiano Royal 1939   63
     Access to the action   64
   Eavestaff Minipiano Royal 1957   66
     Access to the action   66
   Allison upright 1939   68
     Access to the action   69
   Kemble Minx upright piano 1959   70
     Access for repair   70
     Access to the pedal mechanism   70

5. UNUSUAL AND SPECIALIST INSTRUMENTS   73
   F. Menzel upright 1890   73
   Pleyel Wolf Lyon grand 1902   75
   Lindner upright 1970   77
   Marshall & Rose grand 1936   78
   Blüthner upright 1882 tail-less jack   80
   Bord upright 1894   81
   Gustaf Weischel, Elberhudt   84
   Haake grand 1890   87
   Julius Blüthner Aliquot Patent grand 1913   90
   Blüthner 'Hopper' action   91

6. PIANOS WITH PROBLEMS FROM NEW   95
   Alexander Herrmann upright 1970   95
   Challen upright 1976   97
   Daewoo Royalle grand 1988 Renner Action   100
   Kemble upright 1988   102
   Bentley upright 1964 Richard Check Action   103
   Willhelm Steinman upright 1987   106

	Daewood upright new 1990	108
	Starr upright 1987	110
7.	HUMIDITY	113
	Sames upright 1910	113
	Waddington upright 1920	114
	Monington & Weston upright 1930	115
	Bechstein upright 1900	115
	Steinway grand Model B 1971	116
	Yamaha upright Model U1A 1986	117
	Bechstein upright 1882	118
	Bechstein upright 1905	120
8	TUNING: ADDING THE PROFESSIONAL TOUCH	123
	Self-inflicted problems	123
	Design problems within the bearing	125
	Bass tuning problems	126
	Treble tuning problems	128
	The tuning wedge	129
	Sympathetic vibration	133
	Use of the small tuning head	138
	Electronic tuning aids	139
	Playing-to-tuning ratio	142
	Keyboard temperaments	144
	Equal temperament	144
	The interval	145
	Search the bearing for error	149
	The laws of octave balance	150
	Speed test	153
	Octave stretch	153
	Hearing focus	153
	Bass tuning	154
	Pythagorean tuning	155
	Pitch	156
	Earliest Pythagorean intonation	157
	Chasing the wolf	159
	Meantone temperaments	161
	Scale (1)	162
	Scale (3)	166
	Scale (2)	169
	Irregular meantone temperament	170
	Temperament Ordinaire	172

9.	SETTING UP THE BUSINESS	175
	Presentation	175
	Customer relations	178
	Range of tools carried	190
	Castors and grand trolley frames	191
	Tuning charges	193
	Area of operation	195
	Cost of operation	196
	Loss of earnings insurance	197
	Medical cover	197
	Retirement pension payments	198
	Public and products liability insurance	198
	Telephone & answerphone	198
	Advertising	199
	Secretarial time	199
	Accountancy	200
	Income tax	201
	Holidays	201
	Peaks & troughs	202
Bibliography		204
Index		207

INTRODUCTION

Methods of training tuner technicians have changed radically in recent years and have resulted in a serious gap in their knowledge. My own apprentices, on returning from college to me for a period of industrial experience, wanted to take the piano apart to find out what was the matter with it, when the answer to the problem was staring them in the face. Students from other colleges have approached me with similar problems and their comments have always been, 'Why were we not shown how to diagnose problems like this at college?' The answer seems to lie in the fact that college training is directed to 'How to do' rather than 'How to find out what has to be done'.

The book is intended to be used by tuner technicians who have attended a recognised course of instruction and have attained the degree of proficiency required to complete the majority of repairs that are liable to be met with in the course of a normal day's work.

Those involved in the repair, playing, buying or selling of instruments will also find information of considerable value and interest. Pianists are very often asked to pronounce upon the merits of various instruments just because they are capable of playing one, their training having contained little or no technical knowledge apart from the handling of the instrument from the musical aspect of the profession.

The collection of problems is by no means complete, and is not intended to be the answer to all situations but is a start in the right direction.

FOREWORD

In my own apprenticeship days, it was not considered necessary to obtain external qualifications, providing that one's time was served with a firm of repute. James Smith & Sons (Music Sellers) Ltd., 'Piano Makers by Special Appointment to the late Queen Victoria', were just such an organisation.

The prestige of belonging to such an organisation was all that was necessary to indicate both the status and the standards which they required and maintained. Employing a staff of eight tuners and as many repairers and polishers, they were probably the only piano makers in the north-west of England carrying the Royal Warrant. Problems are always present in the servicing of pianos and if there was anything I did not understand, or a problem I could not solve, there was always someone at the shop or the factory from whom advice could be obtained.

Today, however, music houses in general do not employ tuners but sub-contract the work out to students fresh from college. The Piano Tuners' Association hold regular seminars for the further training of qualified students and these are most valuable but at the moment seem to be the only forum for problem solving. Unfortunately, they take place some considerable distance from the main body of tuner technicians and involve them in heavy expense in attending.

In the absence of any such help, the following collection of problems is presented in the hope that it will provide answers to many of the questions most frequently asked.

The pre-tuning chat

Tuning, the adjustment of touch and tone to a particular pianist's requirements or any repair must come only after the instrument has been inspected and an assessment made of its condition, quality and suitability for the pianist's requirements. The initial chat with the customer to find out what will be required of the instrument is time well spent. This can be done whilst taking off the panels and assessing the condition of the internal parts.

It is not the slightest use tuning a poor quality instrument, if it is required for a pianist of grade 6 or 7 standard. No amount of tuning will make a vertically strung overdamper instrument suitable for this grade.

Only the best is good enough for a child to learn on
Not as is generally thought by non-musicians, any old thing as long as it works.

'Piano for sale, suitable for a learner.' Adverts such as this generally indicate an ancient and decrepit instrument of dubious manufacture in need of considerable and expensive attention. Many instruments are bought by unsuspecting purchasers in the hope that it will survive long enough to see their child through the initial stages of learning. It is the job of the tuner during the course of the initial chat to establish whether the instrument can provide this.

Although many pupils are forced by financial circumstances to learn on unsuitable instruments, as long as the customer is made aware of the deficiencies which can't be overcome by any tuning or repair and is willing to allow one to proceed, it is safe to do so, although, as will be established later, this can still cause problems.

As progress is made through the book, some indication as to the possibilities of instruments of varying quality will emerge and even tuners without musical knowledge will be enabled to make evaluations accordingly.

The diagnostic approach

Most calls to service are prompted by piano problems in some form or another. If it is just a tuning problem or minor repair, this will probably be well within the scope of the average tuner to correct but problems of touch, tone, extraneous vibrations, etc. sometimes have to be traced by a diagnostic procedure which has to look far beyond the immediately obvious to reach the root cause.

This procedure will be gone through in Chapter 1, as it will throughout the book, and the general rules will emerge.

Learning to see

This is something which becomes second nature to the experienced tuner, is completed in seconds, is done before any consideration is given to tuning or repair, and involves taking in details of the following:

- signs of moth or mouse damage, damp;
- action, keys and strings which don't look right;
- items which are at odds with the rest of the instrument;
- items which have been renewed or replaced;
- signs of cracking of the wrestplank;
- wear and tear.

All of the above things can be seen and taken into consideraiton before the instrument is touched. In detailing my thought process in approaching each problem, I hope to guide the reader through the diagnostic process.

ACKNOWLEDGEMENTS

My grateful thanks are due to my many customers, past and present, who have allowed me to disembowel their pianos and rearrange their furniture in order to obtain the necessary illustrations for this book. Obviously, all names have been changed but if you recognise yourself, Mrs Brown, I won't tell anyone, if you don't.

My thanks are also due to tuner/technicians, dealers far and near, who have invited me to assist with their problems. Without them, many interesting, complicated and difficult problems could not have been collected.

My long-suffering and patient wife, who became used to me charging off at all hours of the day and night in order to see yet another interesting problem, is now deceased but without her love and support the fourteen years spent in the collection of information might never have been completed.

Special thanks are due to Mrs Joan Booth (no relation) for the loan of her spinet to try out all of the early music forms, and to Mr & Mrs F. Cairns whose helpful advice and constructive criticisms have been instrumental in turning a collection of information into a readable book.

No book dealing with piano problems will ever be completed as there is a never-ending stream of new problems manifesting themselves all the time. Just when I am in danger of feeling that I have seen it all before, a problem comes which taxes both my patience and my experience. The answer, though, is just around the next corner and it is by not giving up that the solution is found.

1

CASE HISTORIES

This chapter is a random collection of case histories designed to show

- the diagnostic approach to fault finding;
- the importance of the pre-tuning chat;
- the art of seeing.

Split wrestplank

In this first case, I was called to inspect an instrument previously tuned by someone else with whom the owner was dissatisfied. The first thing I do in cases like this is to listen to the instrument overall, trying the notes from top to bottom, listening for any sign of rapid deterioration of the previous tuning.

In general, the tuning was quite good but my attention was drawn visually to the two replacement strings of the base singles.

Listening to the strings in this area, I found and marked with chalk the ones which were more out of tune than could be expected considering that it had only recently been tuned. The strings marked were in line, and an attempt to tune them confirmed, by the looseness of the wrestpins, that the wrestplank was cracked in line with the wrestpin holes. It must be remembered that the grain of the second lamination of the wrestplank runs horizontally and it is this which has cracked.

The fitting of

- Larger wrestpins can extend the split further.
- Longer wrestpins can sometimes bridge the gap.
- The sleeving of wrestpins can also extend the split.

It is obvious that the strings which have been replaced had been

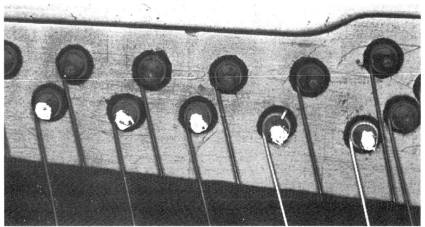

Fig. 1

broken by excessive tuning due to the loose wrestpins and by age. In both cases, the original wrestpins have been used.

The tuning of this instrument should not have been attempted due to its generally poor condition and the fact that during the course of the initial chat, it transpired that it was required for a pianist who was already at grade 4 standard. Although it was overstrung, its age and condition was such that reconditioning could not be justified as the outlay would be more than its value on completion.

Students under training are sometimes shown some very complicated repairs of which, when completed, they are justifiably proud but at the same time, they are not shown how to cost the work involved. In the real world outside college, every minute has to be costed in order to survive. Students' time comes cheap.

UNLESS THE COMPLETED REPAIR enhances the value of the instrument to a point greater than the charge made, there is no point in doing it.

When estimates are made up, the total cost will often be greater than the amount needed to purchase another instrument of equal quality which does not have any problems. This instrument eventually ended up in the saleroom, and the cost of reconditioning together with the amount received from the sale put towards the purchase of a new instrument. Good advice from the tuner and a happy outcome for the pianist.

Advice such as this cannot be given without charge and although the instrument was not tuned, the equivalent tuning charge was made for the time involved.

Fig. 2

Broken hammer shanks

The Chappell & Co upright (see Figure 2), was of pre-First World War vintage. Due to its large dimensions, it had an excellent tone and was typical of better-quality instruments of the period, except in one respect.

The illustration shows that it has suffered many hammer shank breakages in the past. The original shanks are made of red cedar wood, and on the face of it this timber has all the characteristics of beech, namely, straight close grain, cheaper and in more plentiful supply. Many manufacturers chose it as alternative to beech but although cedar wood shingles on the roof of a house will last for a hundred years, indoors and inside a piano the timber dries out and becomes exceedingly brittle.

It can be seen that odd replacements have taken place, but in cases like this it is more expensive to replace them as they break than to replace the lot.

On this occasion, the hammer butts were made of beech so, due to the excellent quality of the instrument, all the shanks were replaced, new heads fitted and the rest of the action was overhauled. If the hammer butts had been made of red cedar wood, it would have meant replacing them as well; it is not the slightest use putting new beech shanks into old red cedar butts.

Although the balance hammer shanks were of red cedar wood, they have never caused any problem within my experience.

The cost of overhaul considerably enhanced the value of the instrument and extended its life for a great many years. Although no consideration was given by the owner to its resale value, on completion this value had doubled.

Blüthner grand Ser. No 119130 *circa* **1934**

The call to service this instrument was prompted by the problem of **sticking notes**. On opening it up, I found that the treble action standard was shattered. It was tied up with piano wire and was wobbling about all over the place. The pre-tuning chat revealed that a previous tuner had been called because of the sticking notes and on leaving had made no improvement. The action standard was pointed out as the obvious cause of the problem. A new one was ordered and duly fitted.

On completion, although there was appreciable improvement, there were still many sticking notes. The feel at the touch was as if something was stopping the hammers from rising. There was very little room between the hammer tail and the rest rail but it looked to be enough. Several of the hammers were removed and at first there seemed to be nothing wrong with them, and it was only when a number were removed and I could see behind the hammer rest rail that the cause was revealed, as was the reason for the broken action standard.

At some previous date, the action had been fitted with new hammer heads and so much glue had been used that globules of glue had run down the underside of the hammer heads as in the illustration. When the reconditioning had been completed, this had caused no problem but as time passed and the hammer rest baize became indented, the angle of the hammer tail increased and the glue was catching against the rear of the rest rail.

The previous tuner had attempted to bend the rest rail supports to move the rail forward and had only succeeded in breaking the action standard. On this model, the rest rail supports are not intended to be bent.

Removing all of the hammers, removing the excess globules of glue and replacing the rest rail baize cured the sticking notes permanently. This is what is meant by getting down to the root cause of a problem. The previous tuner had tried to cure it without finding the cause.

Over-enthusiastic use of glue
The over-enthusiastic use of glue can cause problems and must be avoided as much as possible. In general, the problems are not to the tuner or repairer but show up from the pianistic point of view.

Fig. 3

Many calls are made upon my services, not as a tuner but for my ability to relate a pianistic problem into tuning or repair terms. Such was the case with the owner of a Haake grand, who called me to see if I felt the same way about the touch of the instrument that he did.

The instrument was *circa* 1903 and had been bought as a fully reconditioned model. An examination of the action, case and strings showed a first-class job and I remarked upon the excellent quality, something I see all too infrequently. When I started to play it however, every time the hammers returned to the check, a shock-wave travelled back along the key jolting my fingers, and I agreed with the owner that there was a strange feel to the touch, although at the moment, I could not trace the reason.

On removing the action I discovered that the checks had been recovered and in the process so much glue had been used to glue the cushion beneath the doeskin to the wood, and the same amount to glue the doeskin to the check and cushion, that by now it had become a solid mass and hard as a rock. In order to remove them in the workshop, I had to soak them in boiling water in order to melt the glue.

There are three points to which glue should be applied. After fitting the cushion with only the amount of glue to keep it in contact with the wooden check point A, the doeskin should be attached at the top point B and left to dry as the others are glued on. By the time this is done, it only remains to pull the covering down and glue into place at point C. NOTHING IN BETWEEN.

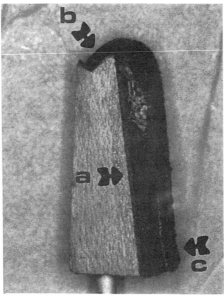
Fig. 4

The average non-pianistic tuner/technician sometimes lacks the delicacy of touch associated with pianists and although it is not possible to describe the feel of every instrument, every endeavour must be made to cultivate a sense of touch in order to trace faults such as this.

Signs of damp
Of the many causes of sticking notes, one of the most common is damp. Fortunately, with the advent of central heating in most homes, this is not now quite so common.

What does damp look like? The typical signs of mildew are shown in Figure 5. The white fuzz is quite distinctive and when it dries out, these marks go black, looking like spots of rain.

Tight bushings are quite a common reaction to damp, and need to be recentered in the usual way, but key and overdamper weights can also corrode and swell. In the case of key and overdamper weights, they can either swell sideways, catching on their neighbours, or they can expand, splitting the keys in the process. Many older instruments have had to be reweighted and the keys repaired due to this problem.

Grand pianos sometimes have key weights at the front, in the region of the sharps, which swell. I carry a small, thin file, slightly thicker than a nail file, which fits between the keys to release them if the odd one catches.

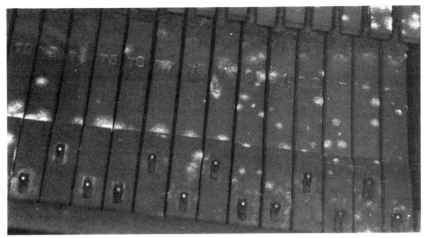
Fig. 5

Signs of mice in upright pianos
Mouse damage causes many sticking notes and is becoming more common, especially in school halls where there is canteen food about. On upright instruments, the signs are always visible on the hammer tails at bass and treble ends and also at bass and treble ends of the keyboard.

On overstrung instruments, these signs appear in the middle of the keyboard at the central break; this is where they climb up and down and is one of the most obvious places to look. All of the mouse droppings must be cleaned from the action to avoid the necessity of return calls to service due to the ones which stick to the central parts only to drop down between the keys at a later date, causing the notes to stick.

Take all of the keys out, cleaning the droppings from the sides, and clean the key bed. The region of the pedals will probably be strewn with litter which must be removed, altogether a time-consuming task.

Mouse damage in grand pianos
Once again, the first sign of mouse damage in grand pianos is sticking notes and this was the case with a grand piano to which I was recalled. The instrument was on regular contract for tuning and had been tuned only a month previously.

On removing the fall, the first evidence was on the top of the keys, which showed urine marks and droppings; the key bed, at bass and treble ends showed similar evidence. On removing the action, there was evidence of their body marks against the underside of the hammers and immediately below, in the well beneath the keys, the start of a nest. The droppings had stuck to various parts of the action and if not cleaned out completely would

Fig. 6

Fig. 7

invariably drop down between the keys at a later date, making them stick.

The action was separated from the key frame in order to clean all parts. The keys could then be lifted out and the sides cleaned, as could the top of the key frame.

(REMEMBER WHEN DOING THIS that any spacers fitted between the action standards and the key frame must be carefully kept in place, the screw which has been removed being replaced in its socket holding the spacers ready for reassembly.)

The well beneath the keys was then cleaned and the action replaced.

The owner assured me that the invaders had been caught and in spite of this, a month later, more sticking notes proved that there were more about and further measures had to be taken to get rid of them.

These are field mice, which come into houses in the autumn when there is a cold snap and can do an immense amount of damage in a relatively short time.

Many years ago, I tuned the pianos at a Church of England junior school in a village near to me a week before the October half term. When the school reassembled two weeks later, I received a call to say that the piano in the hall did not work. I offered to call on my way home that same day. On being shown into the hall, I tried the keys and to my surprise, they would not depress enough to work the action. My first reaction was that something had been dropped down inside and was resting on the keys. Removing the top door, I was met with what appeared to be a large ball of fluff, and there were mouse droppings everywhere. Beneath the keys, there was a solid mass of paper, chewed up material, wood shavings and droppings which took two hours to clean out.

There had been a cold snap during the holiday and the field mice had smelled the gerbils which the children had been keeping in cages in the hall and decided to join them for the winter. Needless to say, the lady teachers beat a hasty retreat, refusing to play the piano until the local Pest Control Officer had assured them that he had effectively rid the school of the unwanted invaders.

Charges for work such as this must be made at the standard tuning rate. The time taken can vary from between one and two hours according to the amount of damage encountered.

Warped lock rail or key slip

The insertion of a lock on an upright piano does weaken the lock rail in the centre, and in conditions of central heating the rail can warp inwards, catching against the key fronts.

Calls to service sticking notes in the centre of the keyboard which on arrival do not stick, are usually caused by the heat being switched off at the time that the tuner calls. A gentle pressure against the lock in the direction of the keys usually makes the notes stick again.

Fig. 8

The cure for the upright is to remove a key in the region of the lock, wedging the rail out from the key bed by tapping in a sliver of hammer shank, cutting it off level with the key frame. It is sometimes necessary to fit a screw between the bat pins to the rear of the wedge in order to give it something to bear against. REMEMBER to mark the key above with the sign of a screw for future reference in case the key frame needs to be removed.

In cases where the lock rail is an integral part of the casework and is not removable, it is sometimes necessary to lift the key frame and reposition it further to the rear. In most cases, this movement is very small and does not impair the working of the keys and action.

New York Steinway grand Model O Ser. No. 237243 *circa* **1921**
Length 70" (178 cm)

This reconditioned instrument was bought from one of our local music houses in 1986. The owner had great **difficulty in getting it tuned** to her satisfaction, complaining that the **pedals were also very noisy**. After being told by the tuner that all Steinway pedals on this model were like this, she accepted the situation, feeling that she must learn to control them better.

A visit from her sister who was also an excellent pianist and who complained of the same things prompted them to call me in to give an opinion.

The tuning was very bad and on asking her to demonstrate the pedals, they were indeed making an alarming noise. Assuring her that Steinway pedal actions should not be heard, I opened up the instrument. Everything inside was perfect.

Removing the pedal lyre, the problem was obvious. The instrument finish had been done in high gloss polyester; it had been sent to our local finisher and assembled on return without examination. All of the pedal felts had been covered in polyester as had the lyre rod bushings, and were as hard as concrete.

The felts had to be removed with a hammer and chisel, the bushings being removed with a large drill. After refelting and rebushing, the pedals could not be heard.

Commencing the tuning, I discovered the reason for the failure of previous tuners to tune to the owner's satisfaction. In the reconditioning process, the frame had been resprayed, together with the agraffes, and the paint was still blocking the movement of the strings through them. The strings jumped alarmingly and although not impossible to tune, were very difficult to handle. As soon as the agraffes ceased in the treble, the tuning became easier. The situation was not improved by the fact that oversized wrestpins had been fitted and some of the new strings were now touching each other.

There was nothing I could do to ease this situation, except to assure her that by the time I had tuned it three or four times, I felt that I could improve the tuning considerably. When both string and instrument stability were established, the tuning improved and although still extremely difficult to tune, now gives excellent service. Problems such as this affect many music houses who have failed to train and ensure themselves of a continuity of experienced personnel. The music house concerned did pay for the work involved but suffered a considerable deterioration in customer relations as a result.

Steinway upright Ser. No. 79784 *circa* 1894

Height 49" (124.5 cm). Width 61" (154 cm). Depth $27^1/_2$" (70 cm). My call to service this excellent Steinway was prompted by the fact that several tuners had **failed to make it stay in tune for very long**. When apprised of the fact, I stated that of all pianos on the market today, Steinway instruments should stay in tune better than ninety per cent of the rest.

Fig. 9

Examination of the tuning found it in chaos, but examination of the wrestpins and bridges revealed no obvious reason. Asking how much it was played showed no surges of activity and an average of about five hours per week, which matched the tuning it had received of twice yearly.

The hammers were worn but not enough to account for the poor state of the tuning. The room was very large and the central heating minimal. The instrument was situated away from direct sunlight.

The decision was made to tune it hoping that something would show up. Nothing did.

Arrangements were made to call back in three months instead of six in order to judge the deterioration of my own tuning. I found it had deteriorated far more than I would have expected. In the process of retuning, I noticed that the slots on the screws securing the pressure bar were slightly damaged, which seemed to indicate a previous effort to tighten down the pressure bar.

Making a probe from a piece of card, I tried to insert it in the top of the pressure bar spacers with no success. They were flat down against the metal of the frame. But when I tried the bottom of the same spacers, the card slid easily beneath, showing that they had lifted from the frame.

All the strings were slackened off and the pressure bar tightened down with a brace and screwdriver bit until the spacers were sitting flat against the frame again. It became obvious that a previous tuner had tried to tighten down the pressure bar with the strings under tension, hence the mangling of the screw slots.

The instrument was returned to pitch, tuned, and has been no trouble since. One does not expect a pressure bar of the size and strength of a Steinway to cause trouble when the size of the average pressure bar is considered, but it shows that even the improbable must not be ignored.

It is possible that during the course of a removal the instrument had been dropped, but as the effect was not immediately noticed, no claim could be made against the removers. I could visualise no other reason for the condition of the pressure bar.

It is not always appreciated that there are many uprights which have no pressure bar at all in the centre section; some Bluthner and Broadwood instruments are constructed in this fashion. The wrestplank is angled back by about 7 degrees and the drilling of the wrestpin holes must be very precise, in order that the strings remain separated and do not bunch together. A close inspection of many of them shows some slight variation in the string spacing but considering that the date of them is *circa* 1895 and in many cases the stringing is the original, there should be no cause for complaint about the design or the craftsmanship.

Steinway upright Model V Ser. No. 285544 *circa* 1937

A 'click' on this model produced a sound which was slightly different from what would be expected if the cushion felts had worn, inasmuch as there was a slight delay when the finger was lifted from the key, before the 'click' sounded. Intrigued, I investigated, removing a hammer.

At the point of attack of the jack upon the notch, a deep ridge had formed in the doeskin. This point coincided exactly with the join of the two felts beneath, and the steep angle of return of the jack to the cushion had pushed the felt downwards until it was tapping through to the wood of the hammer butt.

The delay in the sound of the tapping was caused by the fact that as the cushion had been pushed down, the jack had moved further forward under the notch and as it returned, was catching on the ridge before returning to the cushion with a delayed 'click'.

If asked to repair at some future date, I will allow the jack to sit in a more favourable position beneath the notch by fitting a thinner cushion which will allow the jack to move forward and away from the join between the two felts behind the doeskin.

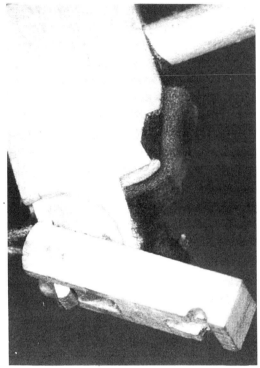

Fig. 10

It is only by following up these slight differences in sound, and being curious about them, that faults like this can be traced and a solution found.

Dagmar upright *circa* 1978

Whilst bearing the Dagmar name, this instrument has the distinct look of another volume maker who will supply instruments without name on request, allowing retailers to put their own name on or a pseudo German name made up by themselves to provide a range of named instruments.

It was sold as a second-hand instrument by a local music house, and the new owner asked her own tuner to call and tune it, complaining to him that the **keys were noisy**. He advised her that they needed completely rebushing and gave her an estimate for so doing.

The size of the estimate prompted her to complain to the vendor that she wanted this matter seen to, or some help toward the cost. Being a reasonable dealer, I was asked to investigate and do what ever was necessary to rectify the problem.

On examination, there was a very strong rattling of the keys but the bushings did not seem to be worn to the extent that they should rattle in this way. Removing the action, to eliminate any rattle from this direction, the keys still rattled. Replacing the action and making a closer inspection of every single key revealed that the rattle was not present in the extreme bass and extreme treble. This would tend to confirm the theory of wear in the bushes, were it not for the additional fact that, rather strangely, the rattle was not present on any of the sharps.

Lifting out a bunch of the central keys, at first I could see nothing, but on turning the keys over, there was a distinct line across the rear of each of the diatonic (white) notes at a point just past the position of the balance pins of the sharps. There was nothing at all on the sharps themselves.

A close inspection of the balance rail at this point showed another line very faintly marked on the edge of what was a very wide balance rail. The distinct line on the underside of the softer wood of the keys did not show up as much on the harder wood of the balance rail.

An inspection of the bass and treble keys revealed no such line but some of the diatonic treble keys had been cut away from a point in advance of the balance pin of the sharps, proving that this problem had been discovered at the factory.

The reason that the problem showed itself in the middle of the keyboard was that the central balance washers had compressed in this area, due to normal wear and tear, allowing the back of the keys to touch the rear of the balance rail as they returned to rest.

The answer was quite simple: 0.3 mm paper balance rail punchings were inserted beneath the balance washers in the centre and the whole keyboard levelled to the point at which it started, and all the rattling ceased. It would have been a very expensive job to have the keys rebushed, only to find that they still rattled.

The charge for two hours' work was considerably less than the cost of rebushing.

AMYL grand *circa* 1928
Length 49" (124.5 cm)

The period between the two world wars was the heyday of the Co-operative Movement, as the largest country-wide super store, supplying everything from food, insurance, clothing furniture and even pianos, all a reasonable prices. Pianos made under the AMYL name, were constructed in great numbers and are still very much in evidence today, in spite of the fact that production ceased at the beginning of the Second World War and did not resume afterwards.

A call to attend to three **sticking notes** was received, and a glance at the keys and hammers was sufficient to indicate the presence of some foreign objects inside the action or in between the keys. A group of three central keys were sticking down, the hammers were also sticking up and could be seen below the strings.

The instrument had been reconditioned, refinished in high gloss polyester and with reasonable care will continue to give satisfactory service for a great many more years.

On removing the music desk, the looseness of the music desk prop, which turned out to be held by only one screw, was sufficient to indicate that the foreign objects would be the missing screws. A close look at the screw holes showed that the additional thickness of the polyester finish had not allowed the screws to return tightly into their original holes. The polyester had flowed into the screw holes and the screws were only holding on to the polyester which had now given way.

Great care must be taken in retrieving the screws from the action. The keys should be levelled and if the hammers do not return to the rest rail, the action should be withdrawn until the hammers reach the wrestplank, then the front of the action should be raised sufficiently to allow the hammers to drop back in order to pass beneath the plank. DO NOT PUSH THE HAMMERS DOWN, as this will surely break the flange or push the head off.

Having retrieved the screws, no attempt should be made to return them to their original holes, which are too short to plug and if longer screws are fitted, they will invariably result in a dome-shaped hump on the front of the music desk panel.

Longer screws may also push a flake off the finish on the front of the panel, which is impossible to replace without complete refinishing.

Another word of warning, this time about trying to shorten longer screws in the absence of the correct replacements: the cut shank of the shortened screw pushes a larger volume of the timber fibres forward and can do the same damage as fitting longer screws.

The more successful way is to mark with a bradawl three new starting holes 1/32nd of an inch (2 cm) lower down the desk panel, refitting screws of the correct size. The cone-shaped hole made by the bradawl is better than a drill for holes of this size and screw length. DO NOT BE TOO HEAVY-HANDED WITH THE BRADAWL.

The distance from the new screw holes to the centre of the hinge pin is sufficient to hide the old screw holes.

Short grands of this type, with their D-type spring and loop action, do not have the length of string, or quality of touch, to provide the pianist with more than a minimum of quality. It is up to the tuner to point this out to prospective purchasers; the music house selling them certainly will not. Having purchased one, the tuner is obliged to explain the reason to pianists who have reached the grade where the problems manifest themselves as they have improved their proficiency.

The proud parent who purchases one of these grands to give their child the best start in music cannot foresee that child growing up into a musician whose discriminating taste at eighteen years of age will find all the limitations of a cosmetic grand. Having said that, the pianist using the above grand is doing grade 8 Associated Boards examination and has found the limitations mentioned above.

Challen upright *circa* 1936

Faint wire sounds are some of the most difficult faults to trace. The slight 'zing' which is heard when a whole chord is struck may not sound at all when single notes are sounded to trace the source. Even when a note appears to generate the vibration, it may not come from that note at all but may be a sympathetic vibration from somewhere else.

Such was the problem with this Challen. When large chords were struck, in the centre of the keyboard, there was a distinct 'zing'. Testing each note separately in the centre failed to trace it; pulling it out from the wall, listening all around it as the owner played loud chords also failed. The only solution was to let the owner play loudly whilst I listened under the keyboard. Suddenly I could hear the 'zing' more clearly, coming from somewhere in the middle section near the central break.

Touching each individual string at the lower end between the long bridge and the hitch pin, I reached a point where my finger reduced it considerably. A closer inspection showed that the tail end of the single eyes on some of the central strings passed over its associated left-hand string, and although there was a felt strip protecting it from the metal of the frame, the left-hand string had over the years sunk into a groove, which allowed the tail of the single eye to vibrate against it when the instrument was played loudly.

Of the seven strings strung in this way, most were either over-

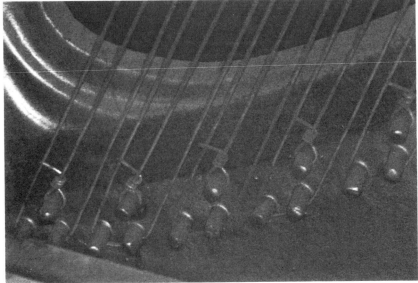

Fig. 11

lapping or at least touching. The solution was to slacken off the middle string on each note, shorten the ends of the offending eyes, retune and test. The result was perfect, and no matter how hard the owner played the vibration could not be repeated.

Students are always instructed to make sure that the tails of single eyes are laid flat against the frame felt or hitch pin washers and do not overlap any other strings. This is a prime example of what happens when this rule is not adhered to.

Loose hammer heads

Loose hammer heads make what can only be described as a cracked sound as they strike the string and there is more than one cause. Most of the glue used on instruments before the Second World War was animal glue (hoof and horn). In centrally heated atmospheres which become very dry, the glue dries out and shrinks, loosening the hammer heads and other joints. Those who have used this glue will remember the shrinkage which occurred in dry conditions when it shrank away from the sides of the pot, falling out in a block if the pot was inverted. This is the same shrinkage which causes the hammer heads to become loose.

The hammer head illustrated was not glued with hoof and horn but one of the more modern bondings. The reason it had become loose was that there had not been enough glue inside the head, which as can be seen is quite clean. Before regluing the head, the old glue must be removed from the shank and from inside

Fig. 12

the head, taking care not to enlarge the shank hole or reduce the size of the shank.

If this is not done, any residue left inside the hammer head and the shelf caused by the old glue on the hammer shank, will, together with the new glue, raise the hammer head above the level of the remainder.

A vertical saw cut in the shank before regluing and a 360–degree turn of the head once it is fitted, will spread the glue evenly around the inside of the head, ensuring a correctly glued joint.

All hammers sent for recovering to pattern must be inspected on return for loose heads, as, due to the differing heights and angles of hammers in the middle section as they approach the central break, some heads have to be removed for this process. Odd hammers which have been replaced at some time have also to be removed, for the same reason.

Balance hammers can give the same cracked sound and should also be inspected. Loose flanges and worn centres give a similar sound and the difference can only be found by experience.

The 'Brasso' man

You may not have one in your area but I have got one in mine. One glance at the overdampers on the instrument illustrated was sufficient to show me that this action had received the attention of our local 'bodger'.

The white substance shown is metal polish and an investigation of the rest of the action showed similar treatment to the hammer and wippen flanges. During the initial chat, it was revealed that the owner had called in the 'tuner'(! ! !), complaining that notes were sticking. She had been assured that although far more than

Fig. 13

a tuning was involved and the action would have to be taken away, it could be effectively repaired.

On its return, the action was demonstrated to her and the instrument tuned. Gradually over the next three months, the sticking notes returned. Repeated telephone calls for service elicited promises to call which did not materialise.

It must be remembered that by undertaking the service of such instruments in this condition, responsibility for their future performance is removed from the previous servicing agency. This is why on this occasion service was refused. When metal polish is used as a lubricant, it does work for a time, just long enough to collect the money and disappear.

After two or three months the notes will stick again and the build-up of the white residue eventually prevents the note working at all. It is not the slightest use after this treatment recentering the parts affected without renewing the flange; recentering will only result in the liquid which has soaked into the wood of the flange working through to the centre again.

This method of easing flanges sticking due to damp originated many years ago, when some of the wooden-framed pianos had actions with hammers strung together on a brass wire along the whole length of the action, the hammer butt being bushed and the continuous metal flange plate holding the brass wire in the centre of the butt as in Figure 14.

Fig. 14

If a hammer in the middle of the action needed attention, due to damp, there was no way that a single hammer could be removed without removing the entire flange plate and unthreading the hammers from the brass wire. Therefore, lubricating the wire and refitting the flange plate was a reasonable alternative.

It must be remembered that in this type of action, the dampers are operated by a wire connected directly to the hammer butt and that the butt is connected to the prolonge (sticker); therefore, in the process of lubrication, with the flange plate removed, movement is restricted to a slight sideways movement. Lubrication was restricted to the use of bear's grease or unsalted beef tallow. Some tuners resorted to the use of the strained liquid from the top of a metal polish can which had been left standing for weeks.

Students are advised – and manufacturers agree – that instruments of this age and type have no musical value and are unfit for use by music students of any grade. This does not stop unsuspecting purchasers buying them because they like the casework and presenting them to their offsprings to practise upon.

Whilst agreeing with the sentiments expressed, in the real world the tuner is faced with either refusing to tune instruments such as these or, having made the owner aware of the servicing and pianistic problems involved, together with the lack of suitability of the instrument to any musical grade, accepting the challenge they offer to their expertise.

For those who spotted the repaired hammers in the illustration, the replacement shank was fitted by breaking the sticker joint and tilting the hammer forward, drilling out the butt carefully and after fitting, regluing the sticker joint. The shank to the right was bound and glued.

In this chapter, we have gone through the diagnostic procedure for a range of instruments from the very good to the very bad, looking, listening and making a diagnosis to get to the root cause of the problem. Before passing on to the problems of specific makes of instruments or actions, we will look at a series of repairs large and small and continue the diagnostic approach to them in Chapter 2.

2

REPAIRS

The small repairs which all students meet in their training will not be covered in this chapter, but a warning is necessary before rushing in to correct all of the items which three years' training has impressed upon minds: STOP, THINK!

- What is the root cause of the problem?
- What will be the effect of respacing hammers or strings?
- Does that hammer flange require recentering, or is it just a loose screw?
- Will correction of hammer travel move the blow to a new position on the hammer head?
- Will re-covered keys, in thicker material, fit into the existing key space?
- Is it a major repair or can it be done on the spot?
- What will be the cost?
- Is the instrument worth it?
- Will the instrument be suitable for its purpose after repair?
- What will the value of the instrument be after repair?

As students are trained, instruments in most cases are completely overhauled and many of the items listed above are not considered, but in practice, they are of prime importance.

Many of the repairs suggested may not fit in with instructions received at college but have been, and must be, tempered by time, cost and – inevitably – the suitability to the pianist's requirements.

Once again, the thought process and diagnostic procedure will be gone through with each repair and the reasons for decisions arrived at explained.

Chatter: cause and effect

Chatter is the double blow experienced when the hammer strikes the string and on return sits on the top of the jack instead of returning to the check. This is usually accompanied by a shallow touch, and is a direct result of wear and tear.

In music, the key signatures of C, G and F are often called the Christian Keys because they are easier to play in than those containing many sharps and flats. All beginners start with them, progressing later to the more difficult keys. In popular music, written for the beginner, the original key signatures are often changed to make them easier to play.

The effect of this can be clearly seen in Figures 15 and 16.

The bushings in the keys C, D, E, F, G, A are very worn but the sharps have received very little wear. This is a very common pattern of wear. As can be seen in close up in Figure 17, the bushings are worn through to the wood. The wood is also scooped out and is typical of the wear associated with 'dog-leg' keys.

The effect at the playing end of the key can be seen by resting a Papps wedge case on the keys. The effect from the pianist's point of view is a reduced touch depth and a double hammer blow, before settling back to the check, when the key is played lightly.

If the bushings are worn to this extent, the hammer rest baize will be indented and the balance washers under the keys compressed to a similar degree. The indenting of hammer rest baizes alone is a frequent cause of sticking notes and will be considered as a separate item later.

In order to correct this common problem of wear, the best place to start is the key bushings. When extreme wear has occurred on one side of the bush, to the extent that the wood is scooped out, it is better to remove the worn side dry with a Swann-Morton surgical scalpel blade No. 11, straightening out the scooped section of the key chase in the process. It is then possible to rebush the key on this side with a thicker bush which will in turn give longer use. If however, all of the keys need rebushing, then this must be done. The keys shown did not need this, and rebushing was completed as described. This now at least makes the keys stand upright, and in some cases (but not in this) is sufficient to eliminate chatter.

If a straight edge is placed across the keys from treble to bass, taking the extreme keys as a measure of the correct key height, due to the fact that they will have received less wear, it will be

REPAIRS

Fig. 15

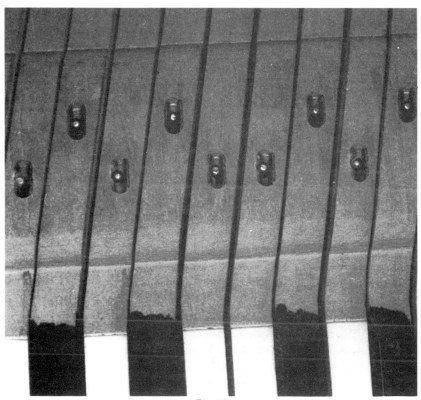

Fig. 16

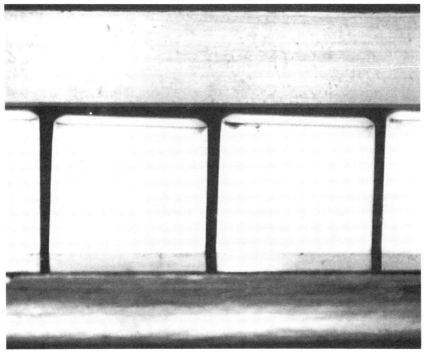

Fig. 17

found that the touch has become scooped out like a saucer, low in the middle and higher towards the edge. The keys must now be levelled with paper punchings under the balance washers until the wear has been replaced.

The hammer rest baize shown was of thick, poor quality material with the consistency of cotton wool, and it was replaced. If the baize is only slightly indented, a decision has to be made as to the method of replacing the wear.

Once again, taking the treble hammers as our measurement of the original blow, this can be done with a card cut to fit between the hammer head and the strings. By placing this against the strings in the middle section, the forward adjustment of the hammers can be gauged. For the bass section, measurement must be taken from the bass singles as these hammers are often at a different level due to the overstringing.

The rest baize can be packed forward in the central area of wear

Fig. 18

by placing a piece of felt such as a nameboard felt between the rest baize and the half blow rail, of sufficient length to cover the worn area, leaving the fine adjustments to the pilot screws or capstans. The existing baize behind the hammer shanks, having already packed down, will last much longer and will pack down less than the new one fitted to the instrument in question.

Do not expect the hammers to be adjustable to an exact straight line against the rest, as the sharps will be worn into differing levels. In practice however, this usually does not matter a great deal. If it looks too uneven, than the rest baize must be replaced.

The before-and-after look of an instrument adjusted in this way is illustrated in Figure 20 and is acceptable.

When only small adjustments are necessary, another method is to place four spaces at intervals between the half blow rail and the action rail (beam), taking up the difference by means of the pilot screws or capstans. Large adjustments must not be accomplished in this way.

IN ALL CASES, PILOTS OR CAPSTANS MUST BE ADJUSTED CORRECTLY.

This adjustment is very often made bringing the jack too close under the notch of the hammer and is in itself a cause of sticking notes. There should be a gap the thickness of a playing card between the top of the jack and the notch of the hammer. To test this, it should be possible to run a finger along the rear of the keys in the region of the capstans from bass to treble without any of the hammers moving in the slightest degree.

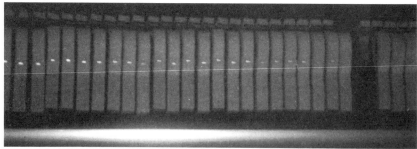

Fig. 19

Fig. 20

In all cases, checks to see if recentering is required should be made on hammer flanges in the central region, especially middle C.

At this point, it might be prudent to mention tapes. If new tapes are fitted, make sure that they are not too short. To test, depress the half blow pedal. The wippens should not be raised by the tapes pulling them upwards. If they do, they are too tight, or the pedal adjustment is too high. When the hammer is at rest, the tape should be slack not tight. If replacement tapes are glued to the balance hammer shank, they should still allow enough slack for the pedal movement.

The practice of replacing the ends on old tapes higher up the tape than the original is frowned upon and is only in order when the shortening effect does not cause any of the above problems.

Time and charges for repairs such as this vary according to the extent of the repair and must be judged accordingly. The job in question was accomplished on the customer's premises, took three hours with a further hour for tuning, the charges being made at standard tuning rates.

Split bass bridge

When I first saw this Broadwood upright, the fall was closed and I thought it was a Challen. On looking up the serial number in the *Pierce Piano Atlas*, I found that in 1932 the Broadwood

Fig. 21

numbers suddenly changed and I wondered if this was another of their excursions into having instruments made in their name by other manufacturers.

Be that as it may, it is an excellent instrument and has had one owner from new. Most of its life has been spent on the African continent, and due to the exceedingly dry atmosphere the animal glue used to fix the bass bridge to the apron had perished, allowing the bridge to lift away at the point of the central break.

The effect was a sudden softening of the tone at the point of changeover from steel to wound strings. There is often a slight change of tone at this point due to the differing strings but the tone here almost disappeared altogether, gradually returning five or six notes lower down.

The crack in the apron and the line of the crack up to this point, was clearly visible through the strings.

An estimate of five hours' work was made and priced at standard tuning rates, plus string breakages, which was the only item which could not be calculated for. This estimate was accepted and arrangements were made to bring along the necessary equipment to tackle the job. On the appointed day, the instrument was stripped and backed down on to a bench trolley.

The strings were

- carefully slackened off, in order not to kink them;
- lifted carefully off the top bridge pins;
- the bottom eyes were threaded on to a screwdriver in rotation; secured with a cork and lifted to one side.

It was then possible to get at the crack. A hacksaw blade was inserted between the bridge and the apron and worked along the

Fig. 22 Fig. 23

line of the crack until the 90–degree crack in the apron was reached. A hardwood veneer the thickness of the hacksaw blade was glued and inserted into the gap, the same glue was worked into the 90–degree crack and the bridge was then clamped to the apron. The excess veneer and glue was cleaned off and left for 48 hours to harden.

In replacing the bass strings, care must be taken to see that they do not kink either at the coil or as they pass over the top bridge pins. At the eye, it is a good idea to give each string half a turn in the direction of the winding, to keep the winding tight, ensuring that the eye is settled against the hitch pin wads or whatever is used to protect them from the metal of the frame, replacing the stringing braid if fitted.

Once the strings were tensioned, the instrument was returned to its upright position, reassembled and the strings brought up to pitch. It was decided to leave the full tuning for a few days for the strings to settle, and the owner was asked to play it as much as possible to take the stretch out of them. No strings were broken, and two years later it is functioning perfectly.

'On premises' repairs such as this can only be accomplished where it is safe from the activities of children and animals or the room can be locked during the 48–hour period on the bench trolley. If this cannot be guaranteed, the more expensive removal to a workshop and return must be estimated for.

Water damage

Vases of flowers and potted plants on the top of upright pianos can cause irreparable damage to the action if spilled or overwatered. Instruments with a central top hinge can let the spillage down inside the action, causing notes to stick.

A call to service with the complaint that the dampers were not working correctly found that the dampers in the middle were not moving at all, holding away from the strings, allowing the sound to linger on. In answer to my questioning, it transpired that a vase of flowers had been spilled over the top and although quickly wiped away the central hinge had allowed the water to seep down into the interior. The flange rail (beam) was directly beneath the hinge and both the hammer and underdamper flanges had received the full benefit of the soaking.

The accident had happened some time ago and the effect on the dampers had only recently shown up. An attempt to remove the affected damper flanges found that the screws had rusted solidly into the flange rail and would not move. It was also impossible to remove the hammers in the same area, due to the same cause.

Sometimes screws in this condition can be removed by heating them with a hot soldering iron, but in this case, as the hammers were working quite well, it was decided to leave them alone, concentrating on the dampers.

The quality of the instrument was poor, and a complete overhaul was out of the question, as the cost would be more than the instrument would be worth on completion. The pianist was of grade 1 standard and as the instrument was overstrung it was decided to get it working again on the understanding that it would only be suitable to grade 3 standard.

It was not possible to apply heat to the damper screws and in cases like this, it is unwise to try to force them loose, in case they break off in the flange rail. Dampers were removed from either side of the affected area until the screws became stubborn; the flanges were then broken off. In all, seven flanges had to be broken off to allow the screws to be removed, fifteen flanges had to be recentered and twelve damper springs had to be renewed.

It was only at this stage that a reasonable estimate could be given as to the charge for service. A loose estimate of three hours' work had been anticipated, including tuning, and an estimate had been given on the understanding that until the full extent of the damage was revealed, it was not possible to be more precise.

Underdamper flanges come in many different shapes and sizes

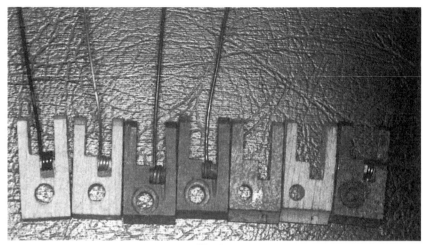

Fig. 24

and care must be taken to replace them with the correct type or a very close match. Looking through my box of flanges produced a variety, some suitable, some not. In Figure 24, the flange on the right is one from the action to which I had fitted a new spring.

The next two flanges, although of slightly different design, had the flange groove and screw hole in the correct place and would replace the old flanges correctly. The next two have the flange groove in the correct place, and although having a larger screw hole and a differing spring fitment would replace the old ones adequately. The screw holes on the final two flanges were in too central a position to be considered as replacements for this action.

In large-scale repair of an action such as this, which requires complete overhaul, send off pattern flanges to the supply house. They may not have the exact replacement flanges, but they will be close enough to be used. For the odd situation, one must be prepared to adapt or even make one from a suitable blank.

The original estimate of three hours proved adequate to cover time, material and tuning.

'Dog-leg' key repair

In rough usage, 'dog-leg' keys split due to the angle of the bushed slot in the key chase, and frequent repairs are necessary.

A quick and effective repair can be completed by gluing with impact adhesive, a thin beech (or similar) veneer across either side of the break, reinforcing the sides which due to the closeness of the slot to the edge of the key chase, would not take a more substantial insert of timber.

Fig. 25

Place the key upright on a flat surface, glue the jagged edges together, making sure that there are no splinters preventing the join meeting. Spread impact adhesive over both sides of the key and allow to become tacky. Spread the same amount of glue over the two pieces of veneer and allow to become tacky.

Place veneers on either side of the key, with the grain running horizontally along the length of the key, remove the keys from either side of the broken key to act as splints, insulate the keys from each other by the use of silicone paper, (computer label paper is ideal for this purpose), bind with string or clamp together and leave to set. This keeps the broken key straight and level, but make sure that the total assembly is flat down against the lower surface. The total assembly should be set by the time the tuning is completed.

Remove from clamp, sand sides, top and bottom. There should be quite enough room for the slightly thicker key to fit back into place without further adjustment. It is extremely difficult to replace the key chase and obtain the exact angle for 'dog-leg' keys. It is easier to rely on the sides to support the key and chase at the correct angle.

If, however, the key chase must be replaced, it is better to make a copy from the old one by placing it above the new key chase block, adjusting it until the two slots coincide, marking off the outside for cutting off.

Before gluing the new key chase to key top, repair as above and, when the key is set, position the new key chase so that when it is placed on the balance pin, the pin is centralised within the slot. Glue, clamp and leave to set.

Whilst keys repaired in this fashion are usable, I usually ask the owner to leave it for 24 hours to allow glue to harden completely before playing.

Grand leg repair

Grand piano legs in a central-heated atmosphere very often develop the 'wobbles'. Most legs are constructed in two parts (some in three), the straight section being dowelled into the leg block; the glue dries out and the joint becomes loose, allowing the instrument to sway alarmingly whilst in use.

In the case of round legs, with the octagonal section, this may be just that the leg requires tightening up in a clockwise direction, this screw method being the method of construction and attachment into the leg block which is part of the body of the instrument.

With square legs, the remedy is usually quite simple, only requiring the removal of the leg cleat (if fitted) and the drilling out of the hardwood wedge from the top of the leg dowel. The dowel can then be pulled out, cleaned and reglued and a new larger wedge fitted.

Such was my anticipation when a leg was brought in to the workshop for repair, but on removing the cleat, to my dismay, there was no dowel visible. This then, would have a blind dowel and it is not the easiest of jobs to separate the leg from the leg block in spite of the looseness of the joint.

There are two methods of approaching this problem. First, to separate the joint enough to saw through the dowel, drilling out the old dowel and fitting a new one. Second, to try to knock the joint apart without damaging the leg block.

It was decided to try the latter and the leg was left leaning against the radiator in the workshop overnight, in order to shrink the joint as much as possible. The following morning, a hard wood wedge was placed against the leg block and given a few sharp blows with a hammer, this being repeated at all four sides of the leg. As the joint slowly separated, it could be seen that there were in fact two dowels, each about 4" (10 cm) long.

When the parts were separated, they were cleaned and all excess glue removed. The most important thing to remember in doing this is to clean out the grooves in the dowels, otherwise it will be difficult to clamp the parts together if the excess glue cannot escape from the dowel holes. Make sure that all old glue from the bottom of the dowel holes is also removed, which is another reason for the leg not clamping together successfully.

After clamping for 24 hours and touching out the marks on the underside of the leg block where the hardwood wedges had damaged the polish, the leg was returned.

The charge for collection, repair and delivery, was two hours at

Fig. 26

standard tuning rates for collection and delivery, plus one hour workshop time.

Coated steel strings

Returning to my workshop one day, I was just in time to prevent my apprentice from cleaning 'corrosion' from a set of bass strings.

The instrument had been brought into the workshop to be completely recentered due to damp and to be repolished. There was a mixture of the usual plated strings and strings which had a green coating on them. The plated strings had rusted and, thinking that the green was also some form of corrosion, he was about to commence cleaning them, until I pointed out to him that at the point of contact with the stringing braid, which had held the damp in contact with the strings, although the plated steel strings had rusted the green coated strings had not.

This was a protective coating which emerged on instruments just after the Second World War, in place of the usual plating and which obviously was still doing its job, in this case better than the plated strings as there was still no sign of rust on them.

Removal of the stringing braid, cleaning of the rusted strings only, followed by the fitting of a new stringing braid, although leaving an uneven look to the strings, made it no worse than its original appearance.

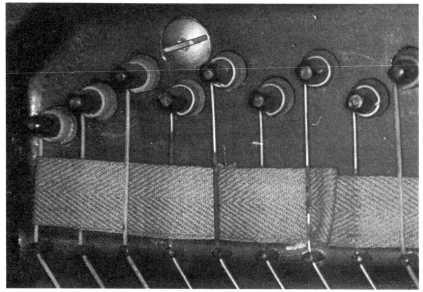
Fig. 27

Synthetic key coverings

The search for a key covering alternative to ivory is nothing new, and has been going on for over a hundred years. Synthetic materials such as celluloid (which used to be made into detachable white collars for Victorian gentlemen) were a frequent sight. Galalith (made from milk) was also extensively used. Some materials such as celluloid yellowed quickly but could be cleaned easily, some could not.

How can the difference be detected?

If a dab of methylated spirit is placed on a celluloid key, the smell of camphor (moth balls) arises; if only the smell of the methylated spirit remains, then the covering is one of the other types.

Many children have been unjustly accused of vandalism on pianos when the pattern of breakages as shown in Figure 28 is discovered. This covering does not yellow as does celluloid but becomes very brittle with age and at some stage, breaks off in the pattern shown. When cleaning such keys in the process of reconditioning, any cracks that appear will indicate that in a very short time they will break off in this manner.

Another type of key covering shows a quite distinctive pattern of discoloration in the area which receives the most wear.

This yellowish/greyish look is indicative of another synthetic key covering which is impossible to clean effectively. A close inspection reveals a crazed surface which, no matter how much

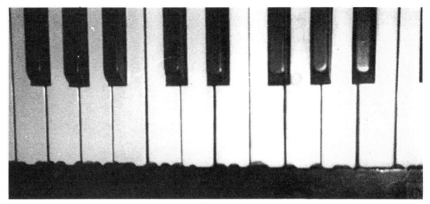

Fig. 28

Fig. 29

cleaning is done, will quickly revert to the same pattern as shown. Over many years, I have tried everything from jeweller's rouge to chemical bleaches without success.

The covering is difficult to remove from the key, but the most effective way I have found is a hot plate or hot iron. BEWARE OF THE FUMES. In removing the covering when hot, care must be taken not to lift any of the timber from the key. Attempting to remove this covering dry causes it to shatter into small pieces and is not really effective.

Modern plastics have yet to be evaluated by time. Their only drawback when used in replacement covering seems to be their additional thickness and their standard size. They are supplied as tops only or tops and fronts combined. The combined form can give problems in the additional thickness of the fronts, which causes them to stick against the lock rail (key slip), necessitating an adjustment of lock rail, or a movement of the key frame to compensate, the lock rail being the easier in most cases.

The additional thickness of the top can cause them to press

Fig. 30

against the nameboard. A thinner nameboard felt or extending the nameboard felt over the top of the bass and treble key blocks can sometimes take this into account, as will reducing the thickness of the nameboard.

In grand pianos, this is more difficult, as the position of the fall is fixed, and especially on Steinway grands there is no room for an upward adjustment of the fall. Therefore, an increase in height of the fall above the keys must be made by taking off an amount equivalent to the increase in thickness of the new covering from the underside of the fall. Thought must be given to these matters BEFORE attempting to replace key coverings or a decision can be made whether to re-cover with

- tops and fronts or tops only, cleaning up the fronts;
- tops and recovering fronts with thin celluloid;
- tops and fronts in celluloid.

Attention must be given to the time and effort of completing any modifications to lock rail, fall or nameboard which is entailed, as this must affect the estimate.

Re-covering keys using tops and fronts combined is successful only when the key is of standard length from tip to the point where the sharp begins.

The repairer of the keys (see Figure 30), was in some considerable difficulty in recovering this keyboard. First, as can be seen, the key is not of standard length and to cut away the uncovered timber between the end of the covering and the sharp, would leave an unacceptable gap between the keys. The second of his problems, lay in the fact that the lock rail was an integral part of the casework and not removable. The added thickness of the key fronts caused the keys to stick against the lock rail when they were pressed down.

The solution adopted by him was to bend the central balance pins backward to move the keys away from the lock rail, this was of course not successful. A subsequent re-covering with oversize key tops, together with fronts re-covered in thin celluloid, would be the answer.

Standard plastic key tops used without fronts give an additional small margin of top coverage equivalent to the thickness of the key front.

Estimates for work such as this will depend largely upon the speed and efficiency with which the work is completed. Whenever the first job is completed, whether it be at college or in the workshop, make sure that it is timed and costed to completion. Money may be lost on the first estimate but the next and subsequent estimates will recoup the amount lost.

Refacing/Reshaping of hammer heads

Refacing/reshaping of hammer heads should only be considered once in the life of a hammer head and this should be done before the head is flattened or cut too badly, let us say, before the hammers in the centre of the keyboard present a surface area to the strings of $3/8''$ (1 cm).

Before this stage is reached, there are other treatments which can prolong the life of hammer heads and serve to improve the tone which hardens as the strings cut grooves in the felt.

A set of hammers was causing one of my customers the problem of hardening of the tone due to wear but had not yet reached the point where the amount of wear justified the complete reshaping of the felts, and the alternative of brushing the heads with a suede brush was suggested.

By this method, the felt is brushed first towards the blow line from the top, then towards the blow line from the bottom, thus reducing the area of blow at the tip (crown), finishing off with a few light strokes to reshape the tip.

The hard flat surface of the suede brush ensures a flat strike surface at the tip and does not round the edges of the felt. The treatment should take about 40–50 minutes. The suede brush should not be used beyond the stage where the wire bristles splay out and would round off the edges of the felt; they are cheap enough to replace, the old ones being saved for use in another situation which we will consider later (see p. 40).

Three instruments I have treated in this way have lasted from three to five years longer according to use, the owners being made aware that the next time the answer would be a complete reshaping of the hammers, which is much more expensive.

There comes a time however, when reshaping has to be done and an assessment has to be made as to whether this has already been done or if the hammers are in their original state.

One of the tell-tale signs of a previous reshaping is to look at the point at which the hammer felt joins the wood on the top of the head. If this has been done previously, there will be a white line at this point showing the amount of felt which has been removed, revealing the glue used to fix the head felt. This line can be ragged and uneven if the job has been done badly and if too much felt has been removed. Look at the extreme treble hammers, where it is more difficult to reshape and the felt is at a minimum thickness.

Some reshaping methods fail to pay attention to the sides of the heads and a glance at them frequently shows a discoloration which does not match the top and bottom of the hammer head.

If the treble hammers are cut through to the wood, the timber of the head can frequently be seen through the felt. If this is the case, it is no use treating the rest of the hammers as there is no way that the tone can be balanced throughout the instrument.

In poor reshaping, there will also be the 'gum-boil' hanging below the head showing where more felt has been removed from the previously dirty top and less from the relatively clean underside.

If however, this has not been the case, and reshaping is considered possible, before commencing, the first thing should be to clean the heads, paying attention to the sides. Clamp the heads together in a vice in fives or sixes, ensuring that they are all level. Brush out the dirt using a suede brush which is no longer suitable for use for the initial treatment of hammers, then clean the sides. If this is not done, the reshaping only grinds the dirt into the felt and necessitates the removal of far more felt than is necessary.

The 'whole-handed' method of reshaping is not recommended, as it is not possible to regulate the pressure evenly over the hammers in the clamp, resulting in badly shaped hammers.

IN ALL CASES OF RESHAPING, A FACE MASK MUST BE WORN.

By far the best method is to cut sandpaper into 1" (2.5 cm) wide strips across the width of the sheet, pulling it from beneath the thumb which is applying just the amount of pressure where it is required. By altering the angle of the hammers in the clamp, this method can be completed from above and below with equal facility. The final reshaping of the strike surface is accomplished

REPAIRS

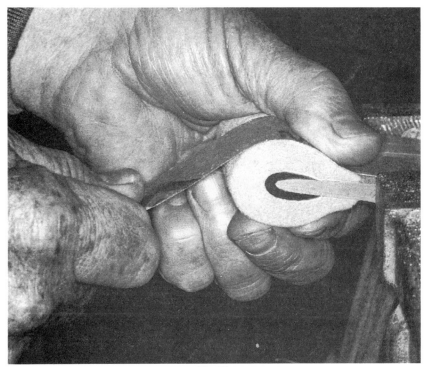

Fig. 31

Fig. 32

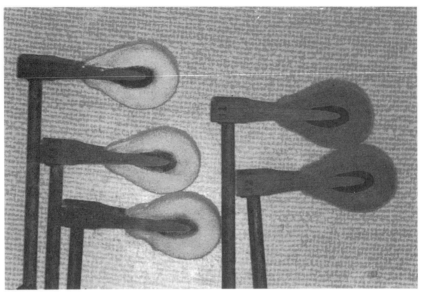

Fig. 33

with finger pressure only. This method allows the final shape of the hammer to mirror the original shape more accurately.

To chalk or not to chalk, that is the question. Many hammer felts even after cleaning and reshaping look old and distressed, and whilst working in a satisfactory manner spoil the look of a completed reconditioning, the colour of the sides of the head not matching the top and bottom. It has been my way to tell my apprentices that our work is like justice, it has to be seen to be done. My local art shop supplied me with a box of 'Playground' chalk about ten years ago and I am still using it. The gasp of pleasure that the customer gives at the sight of the completed work is sufficient to make the extra effort worth while. Look at the before and after in Figure 33 and judge for yourself.

Work such as this is usually part of a larger reconditioning, and charges for time spent must be workshop rates. It cannot be stressed too strongly that time spent at college on repairs should be costed in order that future estimates be assessed accordingly. If records of the length of time it takes to complete a particular job are kept, a realistic estimate can be given. As time passes, experience grows, speed and efficiency increases and charges will become more competitive.

Respacing/Adjustment of hammer travel

In taking my apprentices around with me during their training, it was my custom to leave them with one tuning whilst I carried on to another, returning to check their work later. As I was delayed one day, a very enthusiastic apprentice, having completed his tuning, spent the time correcting the hammer and string spacing.

The tuning was good but when the instrument was played there was a distinct variation in tone from the notes he had adjusted. Odd strings were striking on a softer part of the hammer felt instead of in their normal groove, due to a string adjustment. Other hammers were striking all three strings on a softer part of the hammer head due to the hammer respacing. It took me forty minutes to readjust the hammers and strings so that the hammers were striking in the original grooves cut by the strings.

The only time that adjustments such as this should be made is when the instrument is very new or the hammers have just been reshaped. This is what I mean when I say STOP AND THINK before making adjustments which, however desirable, can have the most undesirable results.

We have looked at various repairs and the dos and don'ts connected with them, in order to stress the theme of thinking and looking beyond the immediately obvious in order to find the root cause. We have considered how far it is possible, reasonable and practical to proceed with repairs taking into account the condition of the instrument and its suitability to the pianist's requirements.

The foregoing applies to all instruments, but it is now time to look at a specific action type and the problems associated with it.

3

'D' TYPE SPRING AND LOOP GRAND PIANO ACTIONS

In this chapter, we are going to consider the problems associated with the range of grand pianos fitted with 'D' type spring and loop actions. These actions were made in their thousands over a great many years and fitted by many well-known makers into their instruments to provide a range of prices. It is not possible to divine the action type from the maker's name because of this.

To all outward appearances, the instruments look the same and sound the same. The case, framing and stringing *are* the same. It is only the action which differs, the more expensive grands containing the better quality roller action.

Calls to service these cheap but popular grands often involve the tuner in much more than the standard tuning. The actions have 'built-in' problems, caused by cheap design and poor quality materials.

Poor quality materials

Cheap manufacture means that everything is at its minimum requirement and of the cheapest materials. It is this which causes tuners the most problems.

All or any combination of

- thin hammer flange rails (beams);
- screws of cheap manufacture with shallow screwdriver slots;
- the attention of tuners with badly maintained screwdrivers;
- hammer flanges without any locating pins

will involve the tuner in spending extra time in regulating, replacement or repair, before tuning can commence.

- Thin flange rails can warp, requiring realignment.
- Screws with mangled screw slots need replacing.

- Flanges without locating pins work loose.
- Screw holes become enlarged due to constant tightening.

It is impossible to replace the screws with longer ones due to the thin flange rail, although the holes can be plugged. A more secure and lasting method is to use screws one size larger in gauge, of the cheese-headed chipboard variety. It is fortunate that the hole in the flange will take this size quite easily.

Anticipating occasions such as this, I carry a box of 500 screws of this size and type, which have an excellent grip, a good screwdriver slot and will not split the flange rail.

If the customer will bear the expense, I replace all of the hammer flange screws, thus obviating this recurring problem and the necessity of expensive extra calls to service between tunings.

Warped key slip or lock rail

The service call stated that the instrument had been removed, that there were some notes in the middle not working and that the pedals did not work. In cases like this, I always make sure that I leave plenty of time to attend to the repair side of things, knowing that I might be faced with a long and time consuming problem.

In situations where pianos are kept in cold damp conditions, the wood collects moisture. If this situation changes, such as the installation of a central heating system, the moisture dries out and warps the key slip, it cannot expand lengthways as it is held by the key cheeks so it warps in the centre either outwards from the keys, in which case it causes no problem or it warps inwards towards them which does cause a problem. Poorly constructed key slips are slotted in place on three dowels which do not hold with sufficient strength to withstand the force of the warp, which is considerable. The key slips which are screwed in with four screws spaced along the length holds much better. The extreme case referred to is the one and only time I have had to resort to thinning down the inside edge of the key slip.

The sticking notes in the middle were caused by a warped key slip, one of the most common problems. In this case, it was cured by making two ivory sliders constructed from two pieces of ivory (heads or tails), saved from old pianos, glued together with the smooth side outwards, waxed and placed in an upright position between the key slip and the key frame.

With the key slip removed, they were placed so that they pressed against the key slip at the two central points at which the slip was screwed into place. When the key slip is replaced, this gives the

'D' TYPE SPRING AND LOOP GRAND PIANO ACTIONS   47

Fig. 34

slider something to bear against, thus preventing the slip catching against the key fronts but still allowing the una corda pedal to work.

In extreme cases of damp, then heat and there is only a wooden dowel to hold the slip in the centre, it is sometimes necessary to thin down the slip where it is pressing against the key fronts in the centre of the keyboard. The slip is held firmly at both ends with no room to expand with damp and can exert a tremendous pressure inwards in the direction of the keys.

There has been only one occasion in my experience when an instrument failed to respond to the first fitting but has responded to the thinning of the slip and restaining to match the polish.

Repairing broken loops

A common problem is the replacement of broken loops. A frequent cause of this is badly adjusted springs. The loop passes through a hole in the head of the jack and if the spring is adjusted slightly to one side, the loop rubs against the inside of the hole, wearing the loop away.

For repairs such as this, I carry a model maker's portable vice with a heavily weighted base with a felt protective covering which can be rested on any suitable flat surface without damage. On this occasion, there were two to be replaced.

A wood drill of suitable size was placed in a pin vice and the old cord drilled out from the rear of the butt. A card template was made from one of the good loops and used to size the replacement loops. A wooden cocktail stick was used to plug and glue the new cord into place. After levelling off, the cushion felt was replaced.

Fig. 35

Fig. 36

Pedal adjustment

Once the action was working, the tuning was completed and the pedals tackled. At first, everything seemed to be in working order and I could not see why the dampers did not lift correctly.

After staring at it for a while, I remembered that I had seen this type of pedal mechanism before, and after a similar removal had had the same problem. The removal men had placed one nut of the lifting (lyre) rod on top of the pedal crank and the other one under the crank, thus clamping it so that it could not move correctly. Figure 36 shows how they should be fitted, both underneath, one acting as a lock nut, nothing on top.

Not all pedal mechanisms are of this type, the majority being of a standard and very well-known design but one must be prepared to encounter the unusual, of which this is only one! The time taken for this work was two and a half hours and charges were made at the standard tuning rates. Workshop repairs are rated cheaper, but must include delivery and return.

Loose flange rails (beams), due to removal

When servicing 'D' type actions, ALL action screws should be checked. If the screws securing the flange rails become loose, they are difficult to reset to the correct distances apart.

Removal damage to these actions is fairly common due to the fact that the body of the instrument is laid on its side for this purpose and there is a tendency for the hammers to slip towards the bass end if the screws are loose. On one occasion, a complaint was received that the una corda did not work and there were many sticking notes after removal. On examination, the notes in the treble were not moving much at all and the una corda pedal was solid, giving no movement.

The fall was removed with great difficulty and on removing the key blocks and key slip, the reason was revealed. The treble leg had been refitted with one of the bolts from the rear leg, which was slightly longer; this raised the action from beneath and stopped it from moving.

Fortunately I had a grand lifter with me, the piano was secured, the offending bolt was removed and measured against a bolt taken from the rear leg. The first bolt was the same size, so this was replaced and the second bolt removed. This was about half an inch (1 cm) shorter and on fitting it to the treble leg, allowed the action, with the correct bolt fitted, to move freely again.

On removing the action, the treble black-leaded action guide had broken away, and had to be reglued back into place. While this was setting, an examination of the action revealed that the hammers and action rail had been jolted out of alignment. THE RELATIVE POSITION OF THESE RAILS to each other is critical to the correct functioning of the action.

Alignment of flange rails (beams)

The action should be placed on a flat even surface before any alignment commences. The position of the hammer flange rail is set, with no room for adjustment and the screws should be tightened up.

The wippen (lever) flange rail with its set-off buttons are secured to three metal action standards by engineering screws and washers (c). These are fitted through a slotted hole in the rail, which allows

Fig. 37

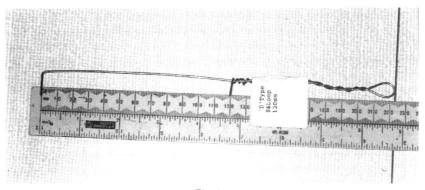

Fig. 38

the rail to move backwards and forwards to obtain the correct adjustment.

At this time, I did not have a caliper to measure the distance between the rails so a wire probe gauge of the correct size was fashioned from a strong piece of wire. The distance from the centre pin of the hammer (a) with relation to the centre pin of its associated wippen (lever) flange (b) should be 120 mm.

The engineering screws (c) were slackened off just enough for the rail to be tapped into the correct position. The treble end was

Fig. 39

adjusted and then screwed down finger tight, followed by the bass end, again just finger tight.

This measurement was then checked at either side of the central break, to see if the rail had warped in the middle. A wire gauge such as this is sometimes easier to use when measuring at the central break, my workshop caliper is a little too large to fit in easily.

A slight tap to get the measurement correct in the centre and a quick check at bass and treble, then the rail was screwed down tight. The remainder of the rail screws were tightened and it only remained then to space the hammers, refit the action and tune.

This call, together with the tuning, took three hours and the charge was made for three hours time at tuning rates. It is not possible to give times for all combinations of problems, but some estimate must be given to the customer before starting.

Variations on a theme

There are many variations of the 'D' type action, and in this model the hammer butt has a long crescent-shaped section to the lower part of the butt where the cushion is normally placed. To cover this, a 1" (2.5 cm) long cushion is provided.

The jack is attached to the key carriage as usual and the set-off rail is in the usual position. But the loop does not pass through the jack; it is attached to the lower end of the crescent-shaped butt and goes around the outside of the jack, looping around the spring.

The adjustment of the spring in a central position is essential to avoid the loop catching against the jack when the key is fully depressed. It can be seen that when the jack is fully extended, the loop is getting narrower, and although there is a felt cushion attached to the rear of the jack to prevent it moving too far back, if this felt gets compressed with use the jack will eventually catch the loop and wear it away.

The hammer flange is of the 'peg' variety and is a variation of the adjustable flange of which we will see more later. The flange bushing is similar to minipiano rear key bushings and there is also an adjustable screw to tighten the bush around the centre pin.

The flange rails on this model are permanently fixed and not movable, being set at the factory, and the shape of the action standards holds both rails permanently in place. There are four of them, which prevents any warping.

The problem with this one however, was not with the action and it is shown here only to illustrate the variation on a theme. The instrument had been reconditioned and the call to service was prompted by the fact that the bass end would not stay in tune.

It was only on looking underneath the wrestplank that the reason could be seen. The wrestpins had been punched down in an effort to make them hold, until the coils of some of the bass strings were resting against the frame above. This had been done without any support beneath the plank and the ends of the wrestpins now protruded through the plank. Fortunately the third lamination had not been damaged and was still in good condition. The strings were also in good condition and it was decided to save them. They were slackened off and the wrestpins carefully removed, slipping the strings out of the wrestpin holes, refitting larger wrestpins through the existing string coils.

An estimate of about three hours was given for the section affected plus the tuning, and the customer was warned that any string breakages must be added to the cost; fortunately, there were none. This is a long and delicate process, if the strings had been in poor condition, this would not have been attemped. No attempt should be made to stabilise wrestpins by punching them down in this manner without support beneath the wrestplank. Even punching one or two down can be all that is needed to push the lower lamination off the plank.

In performing this task, make sure that a wrestpin punch is used: tuning hammers can bend the pins or even break them off.

String coils should at all times be clear of the metal of the frame and if the pins will not hold, they should be replaced.

Tapping down a wrestpin will only firm a doubtful one, one which is holding but there may be some doubt that it will last until the next tuning.

In general, these 'D' type spring and loop actions last well if attended to regularly, constant attention being given to the level of humidity and the above details of servicing. They lack the delicacy of touch and repetition of the roller action, but for the less discerning musicians are quite adequate.

The small physical size of grands in which these actions are fitted, usually 4'6" (137 cm), means that because of the shorter strings, the quality of tone does not satisfy the better musicians, and together with the lack of control of the touch makes them unsuitable for the higher grades beyond grade 6. This does not stop them being used beyond this stage, but it puts the pianist at a disadvantage.

We have considered both the servicing and pianistic aspects of these actions together with the instruments into which they are fitted. It only remains to remind tuners that when giving advice, all of the above must be taken into account and prospective purchasers made aware of the limitations.

4

MINIATURE PIANOS

Of all the various types of instrument produced, the miniature piano gives the tuner/technician most problems.

- Lack of quality sound due to the short strings.
- Difficulty of access to the action.
- Difficulty of tuning.
- Expense of repairing the most trivial of faults.
- Pianistic complaints due to varying touch on the keys.
- Short keyboard (6–octave).
- Extraneous noises from action, keys and attachments.

As pianos became the most popular means of entertainment within the home, there was a general trend to make them smaller. In the 1930s, one of the leaders in the field was Eavestaff, with their 'Minipiano'.

Eavestaff Minipiano *circa* **1936**
Height 33" (84 cm). Width 51" (129.75 cm). Depth 16" (41 cm)

This 6–octave instrument was one of the earliest versions. Earlier models were mostly black, having chromium plated bands but also long chrome light stands at bass and treble ends holding candle bulbs which rattled and buzzed with the slightest over-agitation of the keys. Most ended up minus these accessories due to this problem. Many were later stripped and repolished in mahogany shade, as has been the one illustrated in Figure 40.

Tuning presented many problems; in fact the senior tuners at my firm at this time refused to tune them, considering them to be no more than glorified toys, and many were passed on to me as the youngest qualified tuner. The challenge they presented was considerable.

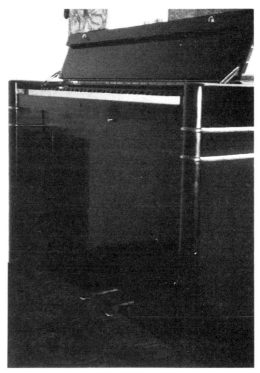

Fig. 40

Firstly, the tuning is effected from the front beneath the keyboard. A plywood flap is pulled out, folding down to reveal the wrestpins. This is not the usual end of the pin with the string wrapped around it, but the rear of the pin which protrudes completely through the very thin wrestplank, the coils being wrapped around what is now the rear of the pin.

These extremely long and thin wrestpins have a square section at either end to fit the tuning key. The most important thing to remember when fitting strings from the rear is that the coils must be made in an ANTICLOCKWISE direction, due to the fact that the tuning is conducted from the keyboard side. This will be seen when we look at the instrument from the rear.

There have been many methods advocated for the tuning of models such as these but all result in considerable disorientation due to repeated operations from either side. It is too easy to wedge the wrong string, ending up with an unnecessary string breakage which, as we will see, is a costly business. It is much better to conduct ALL operations from the keyboard side, in order to maintain orientation and do away with the necessity for a tuning wedge altogether.

Fig. 41

It is necessary, before commencing the tuning, to make sure that each wrestpin is clearly marked with its associated keyboard note. All of the Cs are usually marked with a * and the rest can be marked off from these reference points.

Adequate lighting must be made available beneath the keyboard. Most rooms have a table lamp that is suitable for this purpose. Make sure that you have an extension lead available if the electrical socket is some distance away.

Middle C is the first of the bi-chord strings and from this position downward, the remainder are single strings.

Tuning without a tuning wedge (Bichord instruments only)
The method adopted for the tuning is to first correct the unison of middle C, then strike the tuning fork, counting the number of vibrations required to bring the note into consonance with it. Separate the strings of the unison by the same amount then bring the second string into unison. This may require a slight adjustment by means of the same procedure, but becomes easier with practice.

- C down to G (4th) is a single string and is easy.
- C down to F (5th) is also a single string.
- G up to D (5th) is a double string and must be tuned in the same way as middle C.

Continue through the bearing in this way until completed if all unisons are correct. It is no more difficult making the necessary interval tests than in an ordinary upright. Continue to the bass, all single strings.

To tune above the bearing

- Correct the unison of the F above the bearing.
- Consonance the note from the octave below.
- Correct for the number of vibrations difference.
- Make the fine adjustment to the interval in the same way.

The time taken to tune in this way is offset by the reduced number of strings (107) against the average (222) of a normal upright.

It is unrealistic to expect a high standard of musical quality from instruments of this small physical size.

Smooth progressions of intervals are of the sort that 'Your guess will probably be as good as mine'.

The slightest repair to these instruments, is expensive and when giving estimates, the following procedure must always be taken into account and timed.

To gain access to the action:

- Move the instrument into such a position that access is available from front and rear, this means at least 90 degrees to the wall.
- Unscrew the fall struts at bass and treble ends, which hold it in place as a music desk, allowing it to fold down flat on top.
- Remove the ten screws from the cloth-covered frame at the rear and lift it away.
- Remove the two upward facing screws at bass and treble ends securing the top to the case sides. In some later models, these screws are replaced by metal top door catches.
- Lift top away with an upward and rearward motion to release the V-shaped retention bars at either end. (Remember to reverse this motion when replacing the top, before lowering it into place.)

The keys and 'birdcage' of prolonge wires are now accessible. It is obvious that the action cannot be removed with the 'birdcage' in place, as these are attached to the keys. It is possible to unhook the prolonge wires from the rear of the keys but in practice it is much safer from the point of view of handling the action to remove the leather buttons from beneath the wippens, raising the wires, leaving them attached to the keys.

IT IS MOST IMPORTANT at this stage to remove the pedal rods.

The action is released by removing the two downward facing screws from either end of the key frame. The action will then swivel away from the strings on the engineering screws attaching it to the frame, resting on the case blocks provided for this purpose.

Whilst the action is in this position it is possible to see the construction of the bridges and belly bars (soundboard braces), both on the same side, making the panel below the keyboard the

MINIATURE PIANOS 59

Fig. 42

Fig. 43

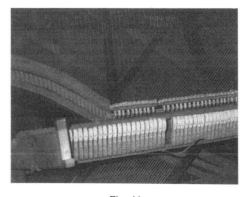

Fig. 44

Fig. 45

soundboard. It is also possible to see the reverse winding of the string coils.

Most repairs are possible with the action in this position. If the action needs to be removed completely, the engineering screws to the rear of the action standards must be removed and the action can then be lifted away from the case. (This is normally a two-person operation.)

On the instrument shown, the problem was a broken damper spring, some broken tapes and a damper felt which had been pushed to one side, probably as a result of the injudicious use of a previous tuner's wedge.

On the single strings, the damper felt is very narrow and is unstitched, which makes it very prone to damage if handled roughly. The width of the standard dampers in the treble can be contrasted with the half width dampers used on the single strings in the bass.

The marks of the single strings against the dampers can be clearly seen, indicating the displacement of the dampers with successive tunings.

It was possible to attend to all of the above problems with the action in this position and it should be noted that the action MUST be moved to this position in order to replace strings. Any attempt to slide them into place with the pedal wedged down will undoubtedly result in damage to the damper felt.

The time taken to gain access to the problem was forty minutes, the repairs took twenty-five minutes, one hour to replace the action and regulate prolonge buttons, followed by one hour to complete the tuning. An expensive operation by any costing standards. A single bass string replacement and tuning can take three hours at ordinary standard tuning rates.

One of the main problems with these models is loose wrestpins, due to the thin wrestplank. To date, I have not found a source of oversize wrestpins of this type, resorting to the insertion of metal sleeves and homemade full and partial sleeves of sandpaper placed rough side to the wood, plain side to the wrestpin.

From the pianistic point of view, all that has been said about the problems of touch and tone apply more in this case than those instruments studied so far. The reduced size, the poor quality tone, coupled with the short keyboard makes them suitable only for the most basic use. Children starting to learn very soon run out of notes when scales are introduced.

Eavestaff Minipiano Ser. No. 6032 *circa* **1937**
Height 36" (91.5 cm). Width 54$^3/_4$" (139 cm). Depth 19$^1/_2$" (47 cm)

As the popularity of these small instruments increased, the quality began to improve and a full keyboard (85–note) was introduced together with a slight increase in height. Access to the action and strings is gained by the removal of the wooden frame cap to the rear of the fall. If this cannot be removed by simply lifting it off, check to see if it is secured by two screws fastening it to the case at the rear; if so, the removal of these will allow the frame cap to be lifted off. This will then allow the fall to be lifted clear of the case.

The nameboard remains in place but allows access to the wrestpins and strings.

The bottom door is no longer the soundboard and can be removed in the usual way. Tuning is then possible without further problems. Once again, unstitched damper felts have been used and care must be exercised in the use of a tuning wedge in order to avoid pushing them to one side.

To gain access to the action:

- Remove the nameboard and the key blocks to expose the keybed to case supports.
- Check the key blocks to see if they have a small locating pin at the front which fits into the lock rail and remember to refit this first before screwing back into place.
- Remove the last two treble keys to reveal the bracket screws.

There are two screws beneath the keybed securing it to the action standards, one at either end. If the keybed is to be removed, this is the stage to do it. Before unscrewing the keybed to case supports at bass and treble ends,

Fig. 46

- unscrew the two screws securing the keybed to the action standards;
- place a support beneath the keybed to prevent the keyboard dropping when the bracket supports are removed;
- when these are released, the keybed can be removed.

In most cases however, this is not necessary and access to the action is gained by:

- First removing the pedal rods.
- Remove the downward facing screws from the brackets at either end of the keyboard.
- Release the action catches.

The action will then swivel away from the strings on the engineering screws, resting on the case blocks provided for this purpose, allowing access to all parts of the action.

Although most instruments of this kind are placed against a wall, great care must be exercised when replacing the action and keys. The positioning of the engineering screws to the rear of the action standards places the point of balance in such a position that the upward and backward movement required to replace the action is sufficient to cause the instrument to topple over backwards.

Some support should be placed to the rear of the instrument to prevent damage to the wall. The instrument illustrated was placed

Fig. 47

across a corner, and a brush stave was placed between the rear and the corner to act as a support.

It is unwise to enlist the aid of the owner to hold the instrument steady, in case of accident.

Whilst being better than the previous model, providing a full keyboard, a larger soundboard and frame, it has not yet reached the quality of touch or tone required at grade 4. The touch is still variable according to the finger placement on the key and these instruments are only suitable up to grade 3.

The time taken to gain access and to replace, was thirty minutes, plus the time taken for any repair.

Eavestaff Minipiano Royal Ser. No. 124478 *circa* **1939**
Height 38$^1/_2$" (98 cm). Width 55" (140 cm). Depth 21" (53.5 cm)

The small increase in height of this model, coupled with an increase in the angle of overstringing, made a great difference to the tone. The increase in depth allowed the keys to be made longer which in turn reduced the variable pressure required in playing. The increase in width was for the benefit of the overstringing. The keyboard was still 85–note. The problem with this instrument, according to the owner, was a soft-sounding area in the middle of the keyboard which was difficult to play.

As soon as I played the offending notes, I recognised the feel of hammers which were catching against the tops of the damper

Fig. 48

heads. Having experienced this problem with three of these models before, I immediately opened it up to feel the underside of the hammer felts and found that they had become detached from the wood of the hammer heads.

To gain access to the interior on this occasion, it is only necessary to lift the front, leaving it supported on its lid support, remove the fall and the bottom door.

Once again, the presence of case blocks to rest the action on was sufficient to indicate how access to the action was gained. The usual engineering screws to the rear of the action standards and some very substantial metal brackets support the keybed to the action.

To gain access to the action:

- Remove the nameboard.
- Remove the key blocks, in order to reveal the keyboard fixing brackets.
- Check to see if there are any locating pins between the blocks and the lock rail, remembering to refit them first on reassembly.
- REMOVE THE PEDAL RODS to prevent breakages as the action is swivelled down.
- Remove the six downward facing screws from the brackets at either end of the keyboard. The keys are now sitting on the support brackets between the action and the keybed.
- Remove the action nuts from the action standards and the action will swivel forward, resting on the case blocks.

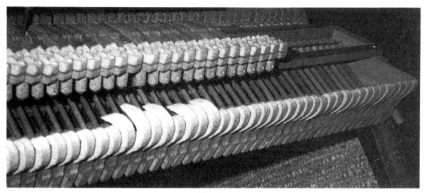

Fig. 49

It was only then, that the extent of the problem was revealed.

As in my previous experiences with this model, my inspection showed that the hammer felt had been fixed to the head with staples which were too short to turn over underneath. With the passage of time, the glue had given way, allowing the felt to spring open, and it was this felt which had been catching against the tops of the associated dampers.

In the short term, the offending felts can be reglued with a hot melt glue gun and clamped. A far more secure method is to drill the heads to take the T-shaped split rivets which will turn over below the heads. Alternatively, the heads can be re-covered or replaced with a new set.

In this case, to facilitate the removal of the hammers, the keyboard was separated from the action as follows:

- Replace action and action standard nuts, leaving the keyboard supported on the angle brackets beneath.
- Place a support beneath the keyboard.
- Unlink the wire prolonges from the rear of the keys, hooking them on to a cord attached to bass and treble action standards. Do not remove them from their original wippens, as they are all of different lengths due to the angle of the action. This saves a great deal of regulation on reassembly.
- Remove the upward facing screws from the angle brackets; the keyboard will still receive some support from them. The keyboard can then be lifted away.

To be conducted with safety, the removal of the action from the case is a two-person operation as follows:

- Release the action nuts.
- Lower the action on to the case supports.

- With one person holding the action, the other must release the engineering screws from bass and treble action standards.
- The action can then be lifted clear of the case.

Gaining access to the action and reassembly only takes as in the previous model thirty minutes. Separating the keys from the action takes considerably longer but is part of a greater repair and must be calculated accordingly.

There are a considerable number of instruments affected in this way, if the range of serial numbers I have encountered with the same problem is anything to go by. Treat these instruments with caution in advising the purchase of a similar model.

The increase in quality of tone and touch has enabled pianists to cope with most musical requirements up to grade 8, whilst, at the same time, realising the deficiencies of smaller instruments.

Eavestaff Minipiano Royal Ser. No. 17020 *circa* **1957**
Height 38" (96.5 cm). Width 57" (145 cm). Depth 19" (48.5 cm)

Whilst bearing the same title as the previous model, there are some differences to be observed. The instrument is the same height, slightly wider and not so deep.

My first impression of this instrument was its high keyboard and I felt awkward playing it until I got used to the higher hand and seating position required. (After getting a higher chair – a kitchen stool – I felt better.) The instrument still has the variable touch as the keys have gone back to the shorter style of the earlier models. The keyboard is now 88-note.

To gain access to the action:

- The frame cap is screwed into the rear of the case and before it can be lifted off, these must be removed.
- Remove fall and bottom door.

It is only now that the essential differences can be seen. The unusually steep angle of the bass strings has necessitated the angling of the action to suit.

The call to service in this case was prompted by the fact that one note in the middle was 'clicking'. Listening to the sound, I felt that it was a cushion felt that had worn through or come loose, and that was my diagnosis.

It was obvious by looking at its closeness to the keyboard that the action could not be removed without first removing the key-

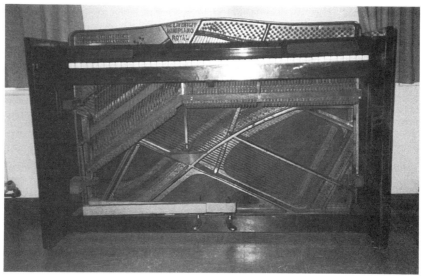

Fig. 50

board. At this stage, it was explained that although the clicking was only a slight problem as far as the playing was concerned, it would take time and expense to correct due to the difficulty of access, and an estimate of one extra hour at tuning rates would be charged in addition to the tuning.

A glance at the action standards showed that they were the usual metal type as in the ordinary overstrung, although much larger, indicating that they are meant to be released by a movement away from the strings. The keyboard must then be removed as follows:

- Unscrew the nameboard from bass and treble key blocks.
- Remove bass and treble key blocks. Check to see if there are any locating pins which fit into the lock rail and remember to replace them first before refitting.
- Remove the upward facing screws from the two wire brackets fixed between the underside of the keyboard and the prolonge (sticker) flange rail (beam).
- Slide the keyboard away from the action, just enough to commence disconnecting the prolonges. Note at this stage that the keyboard is fitted on runners in a similar fashion to the grand music desk.

When disconnecting, PUSH the capstan adjuster at the top of the rear of the key and also the prolonge wire just below the key, out of its V slot. If pressure is applied to the top only, the wire will be bent and the whole keyboard will have to be regulated on refitting. Remember to PULL them back into place on refitting.

When this is done, the keyboard can be slid away from the case and removed. The action, prolonges and prolonge rail can be removed in the same way as an ordinary overstrung action. It is not necessary to remove the pedal rods in this model.

The action will stand on the metal action standards and allow work to be performed quite easily. If necessary, the prolonge flange rail can be detached by the removal of six engineering screws fixing it to the action standards. In this case, it was not considered necessary.

On removing the offending hammer, to my surprise, it looked all right and for a moment, I was at a loss to understand what was causing the click. An inspection of the cushion felt, however, revealed that in a previous reconditioning it had been glued at the rear to the hammer butt and also to the doeskin above, and the sound was coming from the jack striking the glue above on its return to the notch. Here is yet another case of the over-enthusiastic use of glue.

The glue was removed with a scalpel and on refitting the hammer, the click had disappeared. The owner was advised that if one of these cushion felts had been replaced in this fashion, in all probability, the remainder would at some time or other react in the same way. The remainder of the action was in a considerable state of wear, however, and she decided that if this showed up again she would have the action completely overhauled.

The actual time taken was forty-five minutes and the hourly charge at standard rate was made, plus the tuning.

There was some difference in the tonal quality caused by the increased overstringing, but taking into account the high keyboard and the shorter keys, from the pianistic point of view, I did not feel that it was as good as the previous model. Having said that, this instrument was quite adequate for the pianistic requirements of the owner, and if reconditioned would continue to provide for her requirements in the future. The cost of reconditioning would be considerable, but would be much less than the cost of a new one.

Allison upright Ser. No. 56487 *circa* 1939
Height $37^1/_2$" (95.5 cm). Width 53" (135 cm). Depth 19" (48.5 cm)

After removing top and bottom doors, fall and nameboard, once again there is the same indication as with the Eavestaff, that the action is designed to be moved towards the keyboard but that

Fig. 51

there is not enough room for this to be done without first removing the keyboard.

To gain access to the action:

- Remove the key blocks. This avoids any damage to the key cheeks as the keyboard is removed.
- Remove the upward facing screws from the brackets beneath the keyboard at bass and treble ends.
- Lift the keyboard away from the action just enough to commence removing the wire prolonges.
- Lift the prolonges out of their sockets, pushing them out from the V slots in the same way as with the Eavestaff, from above and below.
- Lift the keyboard away.
- Release the action by means of the securing catches. It can then be lifted clear of the case.

It is not necessary to remove the pedal rods before releasing the action but it is easier to remove the damper lift rod and refit it to the action before reassembly.

On this occasion, it was necessary to remove the prolonge rail assembly to make the necessary repair. It was separated from the action by the removal of two screws at both bass and treble ends.

Once again, the time for disassembly, repair and reassembly was forty-five minutes and the hourly charge was made, plus tuning charge.

The keyboard in this model is in a better position but the same problem of varying touch remains due to the short keys. Because

Fig. 52

of the low overall height, the music desk is at keyboard level, holding the music at an unacceptably low level for some pianists.

Kemble Minx upright Ser. No. 109007 *circa* **1959**
Height 35" (89 cm). Width 53" (134.7 cm). Depth 23" (58.6 cm)

To gain access to the action:

- Raise top lid.
- Remove the small top door.
- Remove the fall.
- Remove the bottom door.

My first impression of this instrument was that it would be quite difficult to service, but to my surprise it was remarkably easy. There was considerably more room between the action and the rear of the keys than in other models. The rounded shoulders of the case were held on by metal catches which, when unfastened, allowed them to be detached from the case.

To gain access for repair:

- Remove the screws at bass and treble ends securing the wooden action standard to the case block.
- Release the action catches.

The method of handling the action is important, otherwise one

Fig. 53

can leave a considerable amount of skin on projecting parts of the keys and prolonge mechanism.

- Grasp the action by the hammer rest rail with the thumbs on the outside fingers on the inside, pushing the hammers forward in the process.
- Lever the action away from the strings.
- When the dampers are clear, lift the action upwards clear of the case.

A note about the keys and prolonge mechanism is worth mentioning at this stage. The keyboard is an integral part of the casework and is not removable, as is the prolonge flange rail. Attention can be given to all parts as all screws are readily accessible.

Having dealt with the action problem, the pedal mechanism was making a dreadful noise. There have been many attempts to improve pedal mechanisms over the years but this is not one of the successful ones. The pedal mechanism is shown in Figure 54. To service, the mechanism should be removed as follows:

- Unscrew the pedal hook.
- Unscrew one of the pedal's pivoting blocks and remove.
- Pull the pedal out of the other pedal block.
- Remove the screws from the central pedal rocker block.
- Disconnect the damper lifting rod and remove mechanism from case.

Fig. 54

The action of the mechanism is to flex the steel bar over the head of a dome-headed screw, which screeches loudly in the process. There is no permanent cure for this and in this case the screw and metal bar were heavily waxed. This will no doubt wear through in time but is better than nothing. Oil has been tried, but hard beeswax seems to last longer.

Although access to the action is much easier, the pedals are a distinct and recurring problem. The time taken to repair the pedals and reassemble was thirty minutes, coupled with a small repair to the action, which, if there had been no pedal problem, could have been accomplished without charge. The hourly charge was made at standard tuning rate, plus tuning.

There are many variations on the theme of miniature pianos but the method of access and the things to watch for in the essential differences have been covered in this chapter. Proceed with caution, looking for the next logical step.

Organists seem to like these instruments as they are accustomed to short keyboards, and for hymn playing they are ideal, but sooner or later most pianists find the limitations.

5

UNUSUAL AND SPECIALIST INSTRUMENTS

There have been many variations of piano actions and stringing methods over the years. In this chapter, it is intended to show:

- The means of gaining access.
- Regulation and tuning.
- Forewarning of the problems to be expected.
- The means of avoiding or correcting them.

F. Menzel upright *circa* 1890

This large overstrung instrument is typical of its period, with marquetry and mother of pearl inlay, candle holders and ornate carved case. It is in its original condition, is on pitch and in regular use.

At first glance, there is nothing unusual about it. However, a closer inspection reveals that the dowel prolonge connecting the key to the wippen is unusual in that it goes inside the key to a depth of about $^3/_8$" (1 cm). This is a clue that there is going to be an unusual method of removing the action.

In Figure 55 the pilot screw can be seen inside the key block as can the felt pad on the end of the prolonge. From this, it can be deduced that the action cannot be moved away from the strings without breaking off the prolonges.

If the action securing buttons are released, it will be found that the action will not move. The wooden action standards are securely fastened to a block which is an integral part of the key frame.

In the case in question, a glance at the screws securing the balance rail at the bass and treble ends showed that they had been removed many times before. When I removed them I expected the keyboard and action to slide away from the strings, but nothing happened.

Fig. 55

Fig. 56

The next clue was found near middle C. The mark made by the head of a screw had been burned into the top of one of the keys. On lifting the key, a large screw was revealed in the centre of the balance rail securing it to the keybed. When this was removed, the action, keys, keyblocks and keyframe slid away from the strings, allowing access to all parts of the action.

In reconditioning actions and keys, attention should always be paid to keys marked in this way. If they are removed in the process of cleaning, they should always be replaced for future reference.

The only problem this instrument suffers from is damp, being in a Gospel Hall which is used infrequently. The installation of a Dampp Chaser heater solved the problem admirably. It has been found in practice, that except in the most extreme conditions, this 15–watt heater is quite adequate and can be fitted into upright instruments quite easily. (*Note*: Once installed, the heater must be left switched on permanently, summer and winter. This way, even if we get one of our soggy summers, the atmosphere within the instrument remains constant.)

At this point, I would like to tell of the only occasion when these heaters apparently did not work. Having fitted a heater into an upright at a local church, I received a call some eight weeks later to say that the instrument had made no appreciable progress. On calling to inspect it, I found that indeed, there was little if any change. The vicar assured me that the heater had not been switched off, pointing to the note on the plug asking that it should be left switched on. Other than asking him to continue with the treatment for a further month, I could see no reason for its failure.

As we left the church, the vicar reached up and knocked the main light switch off to extinguish the lights. It was then I realised that in doing so, he had also switched off the heater. There was a large notice above the main light switch reminding church users to switch off before leaving the church and the only time that the heater had been on was when the building was occupied. When the heater was isolated from the main lighting circuit, the instrument swiftly recovered, and ten years later is still in use.

All problems are not necessarily of a technical nature as we can see. Some of the most difficult ones are capable of a very simple solution, the only difficulty is in finding it.

Pleyel Wolf Lyon grand Ser. No. 127267 *circa* 1902
Length 62" (157 cm)

There are several items worthy of note on this grand, not least the spring arrangement.

Instead of a single or double wing spring, a large spiral spring is fitted between the underside of the cradle and the elbow of the jack. At the top, the spring sits on an adjustable dowel and at the bottom, the smaller end sits in a circular hole in the elbow of the jack, in much the same way that the spiral jack spring sits under the tail of an upright jack.

Fig. 57

The top of the cradle is shaped to fit the notch in such a way that it rests against both the cushion and the nicely cushioned doeskin notch. The jack is tensioned by an adjustable screw which is fitted into the dowel, holding the jack against the notch. The hammer butt is shaped in much the same way as an upright butt.

The spring is a most successful arrangement and I have never had to replace one but have removed one to test the ease of service. They are quite easy to remove and replace. The physical size of the spring would mean that they would have to be ordered as a special part.

The working parts are very well insulated from one another, making for a very quiet action. There is an unusual pilot covering attached to a block on each key to the rear of the capstan screw extending forward between the undercarriage and the capstan screw. The red cushion felt fitted to the tail of the jack is also unusual as is the very thick cushion fitted to the check.

There is no drop screw fitted through the flange as in some modern grands but there is a T-shaped drop screw fitted to the underside of the hammer butt, the side view of which can be seen between the end of the carriage and the underside of the brass hammer flange.

The set-off adjusters are fitted through the flange rail in front of the sectional brass flanges. For those not familiar with these flanges, they are in two parts, the top half fitting against the bottom half by means of locating pins. Unlike the long sections of earlier models, these short sections are much more accessible.

Each section of flanges has a continuous wire for the whole of its length. When this is placed in the groove, the top half is screwed down gripping the wire tightly, spacing the hammers accurately. The hammer butt is bushed in the same way that an ordinary hammer flange is today.

Wear in the butts can be rectified a section at a time, but extreme care must be exercised in the removal of the central wire, as it is very easy to pull the bushings out of the butts in the process. If it is decided to rebush all the hammer butts, then the original wire can be replaced, taking care to ream the bushes to the correct size and to beeswax the wire to reduce the drag on the bushings before refitting. If wear is not too great, the bushes can be reamed to the next greater, wire size and a new wire fitted. Don't forget to number the butts before removal, replacing them on the wire in the correct order and to beeswax the wire before refitting them.

THE REPLACEMENT WIRE MUST BE STRAIGHT. To ensure this, cut it to length and roll it between two hardwood blocks until straight. The larger gauges of piano wire are usually sufficient to encompass all sizes likely to be met with before rebushing of the hammer butts is required.

The lyre fitment is unusual inasmuch as the two wooden uprights are push-fitted into two metal blocks and are pinned into position by two metal ring pins pushed through the metal block and the wood until secure. There is of course a metal strut to the rear of the lyre, giving it something to bear against in use. All in all, in spite of its unusual features this instrument

- works effectively;
- has lasted for 90 years;
- has caused no problems;
- is very smooth in its touch;
- is silent in operation;
- has a string length which brings the tonal quality well within the range of most pianistic requirements.

Lindner upright *circa* 1970

It has been my endeavour to show all types of instruments together with their associated problems, from the excellent to the mediocre, but there have been none so bad as the plastic-actioned Lindner uprights.

Assembled in the free trade area of Shannon Airport from parts shipped in from Holland, all of the action and keys were made of parts which were non-standard and impossible to replace with orthodox parts. The hollow plastic keys were of the worst design

I have ever come across. The movement was provided by the constant flexing of a piece of metal which inevitably gave way under the strain and had to be replaced.

This constant replacement was possible when they were readily available, although at a cost in service time to the vendor or to the purchaser when the guarantee ran out.

Even now, I get calls from excited prospective purchasers to say that they have found a modern piano at a very low price, which has only one key not working. On asking the name of this bargain, if the name Lindner or Copic or any of the names used to disguise their origin crops up, I immediately advise against purchase, even if all the notes are working.

As the reputation of these instruments deteriorated, it was inevitable that production ceased, together with the supply of spare parts. This damaged the name of the Dutch firm supplying the parts as it was obvious that they had not used their own name. From then on, I have always treated instruments from this source with suspicion and have never recommended them to prospective purchasers.

When asked to service or tune them, I decline, advising the owners to contact the music house which supplied them, knowing that they too cannot service them. The excuse the owners will receive is that the firm has gone out of business and the parts are unobtainable. This reflects badly on the music house and damages their reputation for selling such poor quality instruments.

A question I am often asked is why I do not sell pianos, and my answer is always the same: I would have to sell instruments in which I had no confidence in order to compete and make a living and my conscience would not allow me to do this, when I can make a living with honesty and integrity in my capacity as a piano tuner and repairer.

Take a look around any local music house and see how many instruments of dubious quality from equally dubious origins make up the majority on view, and judge the standard of the music house accordingly.

Marshall and Rose grand Ser. No. 36367 *circa* 1936
Length 51" (130 cm)

Whenever I see a multiple bridge arrangement such as the one in Figure 58, I know that I have to take great care in removing the action. Not only will the hammer heads be on three different levels, the higher position of the hammer flange rail (beam) will

UNUSUAL AND SPECIALIST INSTRUMENTS

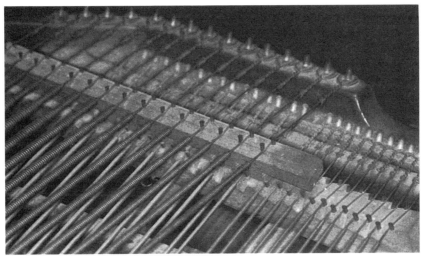

Fig. 58

quite often not pass under the lock/plunger rail and even if it does, the hammer butts certainly will not.

On this instrument, after removing the keyblocks and key slip, the hammer flanges would pass beneath the lock rail but the hammer butts would not, and the rail was removed. The four screws securing it to the metal of the frame allowed the rail to be lifted up revealing the hammer butts. Many bass hammers have been broken by tuners who withdraw actions of this type without taking the necessary precautions.

In most cases, the lower bass hammers being on the highest level will just scrape underneath the wrestplank, but in some, the hammers will not. In cases like this, some makers give instructions on the hammer rest rail for the successful removal without damage. For occasions when instructions are not given, the procedure is given below.

To withdraw the action:

- Withdraw the action until the hammer butts are clear of the wrestplank, but not far enough to catch the hammers on the underside of the wrestplank.
- Raise the front of the action (keys) just enough to allow the hammer heads to drop back slightly.
- Continue to raise the front as the action is withdrawn until the hammer heads are clear of the wrestplank.

To replace the action:

- Hold the front (keys) high enough to allow the hammer heads to pass beneath the wrestplank.
- Lower the action when the heads have passed the far end of the wrestplank and the near side of the wrestplank is approaching the hammer butts.

The pedal spring on this model is fitted in the key well, below the action. It is quite clear of everything and causes no problem as long as one knows where to look for it when waxing to ease a squeak.

The damping in the bass was causing a problem and when I looked, there were two causes. Firstly, the bass single damper heads had splayed out and were catching on one another, a very common problem. In this case, it was compounded by the singular lack of separation between the last two strings. As a result the damper heads were impossible to adjust.

The solution was to fit new damper felts, chamfering the tips of the felt where they were close together to relieve any future effects of splaying.

Blüthner upright Ser. No. 20680 *circa* 1882 tail-less jack

There are a number of upright instruments still in operation with tail-less jacks, which are attached to the wippen by a sprung flange similar to the underdamper flange. The spring sits in a slot at the rear of the jack in a similar way to the slot at the rear of the underdamper body, ensuring contact of the jack with the notch of the hammer butt.

Set-off is achieved by a grub screw adjuster situated to the right and to the rear of the jack which operates a doeskin covered button which slides against the rear of the jack.

The rear of the jack is black leaded and as the key is pressed, the wippen raises the jack which slides against the button. The shape of the rear of the jack provides the set-off, according to the adjustment of the set screw.

In cases of breakage, jacks have to be repaired as replacements are not readily available. Such a repair is illustrated in Figure 81. The new top section has been butt jointed to the existing shaped lower end, glued and dowelled into place.

The keys on this particular model have an adjustable back touch rail instead of the usual individual front touch washers. Although this works effectively, it puts a strain on the key at the balance

Fig. 59

pin, many keys in overstrung models breaking at the central key chase.

Also on this model is the patent brass compression flange. The hammer flanges are made of brass, they are fitted with a grub screw, threaded into the rear section which when tightened serves to clamp the bush tighter, taking up any wear. There is a slot in the brass flange which fits into a locating pin on the flange rail. The flange rail has no groove and the locating pin stops the flange from slipping sideways. A strip of box cloth runs the length of the rail beneath the flange which masks any rattle of the flanges should they become loose.

The slotted heads at the rear of the set-off adjusters can be seen below the hammer flanges. Should spare parts become available for such models, it is worth buying them if any large scale repairs are envisaged, as they can be extremely difficult to replace effectively.

Bord upright Ser. No. 84356 *circa* **1894**
Height 45" (144.5 cm). Width 51" (130 cm). Depth 21" (53.5 cm)

This instrument is regularly tuned, in perfect working order, regularly used and is typical of the touch and tone of the period. The popularity of these small lightweight wooden-framed instruments as opposed to the general run of the larger iron-framed instru-

Fig. 60

Fig. 61

Fig. 62

ments of the same period means that there are still a considerable number of them in use today. Many of them were taken out to Africa and India by soldiers and missionaries because of their light weight and were carried to the most inaccessible places by native porters.

By the time they arrived, I dread to think what they sounded like; there must have been some pretty poor hymn singing.

With humidity and central heating kept to a minimum, they can be kept in working order quite easily. The wrestpins are of the oblong type and a tuning key of the correct shape must be used. On this model, however, the wrestpins are square, probably made during the period of change from oblong to square.

Some makers give instructions for the regulation of the action, displaying it on the hammer rest rail (beam) as follows:

PATENT 114, 1896 REGULATING HOPPER
By this principle, the creaking of the hopper upon the lever and the blocking of the hammer is entirely avoided. Regulate so that the hammer shall be within touching distance of the string and be relieved by a slight blow.

The hopper is shown in side view (see Figure 62), the adjustment is made by turning the adjuster clockwise to reduce the distance of the hammer to the string, and anticlockwise to increase it. The top of the hopper is blackleaded as is the slot in which the spring sits. The spring is fitted completely through the hopper base and is turned up at the rear to prevent it moving to right or left. The vertical slot of the vellum joint is clearly seen and are all in original condition and in working order. Not bad after ninety-six years.

To replace a joint such as this, a vellum saw is required. This saw has teeth but no set, the slot must be as thin as possible. The horizontal slot on the hopper to the left of the vertical piece of vellum is to allow it to be fitted without splitting the hopper. Much of the vellum supplied today is too thick to repair joints such as this but material such as vylene, parchment and even negative film has been used quite successfully in the past.

The secret of a successful repair is to ensure that no glue is present at the junction of the two pieces of wood, where the joint is flexed. Apply glue to the piece of material used and slide it into the slot sideways, cleaning off any excess glue from the side of the wood.

The keys are invariably unbushed but because they are at right angles to the action and strings, show very little wear. If they do get noisy, a slip of silicone paper from the back of my computer

84 PIANOS, PIANO TUNERS AND THEIR PROBLEMS

Fig. 63

labels, glued dull side to the wood, shiny side to the pin, works beautifully. Ingenious, these repairers!

Whilst agreeing that instruments such as this are museum pieces rather than successful musical instruments, the rate with which they are encountered prompts me to give some idea of their repair and regulation.

They are unsuitable for any child to learn on, due to their unusual tone and in most cases result in the child thumping on them to try and make it sound like their music teacher's piano. They are only suitable for those whose interest lies in either antiques or the music of a period in which most instruments were of this construction.

Gustaf Weischel, Elberhudt

Although no date can be found for this instrument, by its case, size and decoration, it is of pre-First World War vintage, having an excellent tone and being in perfect working order. Because of its unusual action, it deserves special mention, as the working and adjustment of it is not easily understood without removing the bass action standard.

UNUSUAL AND SPECIALIST INSTRUMENTS

Fig. 64

There are three actions rails (beams), the top one supporting the hammer and damper flanges, the central one supporting the wippen flanges and the bottom one supporting the prolonge lever flanges. The hammer butt has a pendulum construction beneath the notch (shoulder) and a ledge below this into which fits the angled tip of the jack. The lower end of the pendulum is covered in doeskin.

The jack is of a most unusual shape and it is only when the action is in motion that the reason for this becomes obvious. With the action standard removed, the jack spring can be seen. It is a small spiral spring attached to the lower rear end of the jack between the jack flange and the check lever.

The jack sits on an H-shaped flange attached to the wippen. Between jack and wippen is a lever to which are attached the check, jack spring and check adjusting screw.

When the key is depressed, the hammer moves forward until the set-off is reached, then continues in motion until arrested

Fig. 65

by the check on the underside of the pendulum. It can be seen at this stage that the hollow in the body of the jack is to accept the pendulum in its forward swing.

There is no tape to assist the hammer to overcome the inertia between the blow and its return to the check, but when the key is released the shaped tip of the jack hangs on the ledge below the notch, pulling it back as would a tape in the tape check action. Note that the bottom ledge is also covered in doeskin to minimise any action noise, and that the jack head is blackleaded in the usual fashion.

The damper flanges do not fit to the top of the action rail as in modern instruments but have a central flange with a locating pin to the rear of the flange rail (beam). The body of the damper is bushed and the centre pin is held stationary within the flange.

The spring fits to the flange, resting at the top of the damper on a roll of bushing cloth instead of the usual hollow. The only problem in servicing would be the replacement of any broken

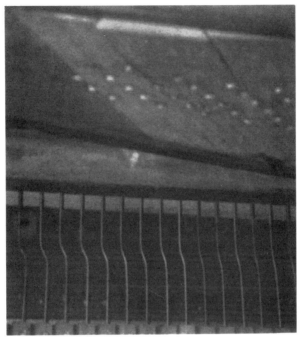
Fig. 66

wooden parts which would have to be made and replacement of the spiral jack spring which would have to be ordered or adapted to take another type of spring. There is very little space between the jack and the check to take an adaptation, and probably a spiral spring fitted between the tail of the jack and the top of the check lever would solve the problem.

Haake grand Ser. No. 30390 *circa* **1890**
Length 59" (150 cm)

This grand was purchased at a house auction. It had obviously been repolished and reconditioned, and the problem complained of was sticking notes and unstable tuning. At first glance, it looked to be a first-class job. The sticking notes were easy to diagnose: on removing the fall, the residue of Christmas party crackers and a pencil were lodged between the keys and the action standard.

The action was taken out to remove the various bits and pieces, and it was then that the reason for the instability was revealed.

In the process of restringing, the wrestpins had been replaced without support beneath the wrestplank and the third lamination of the plank had been split. However, the wrestpins were not too

Fig. 67

Fig. 68

loose and it was decided to leave well alone, concentrating on keeping the central heating down and the humidity up. Luckily, the lady of the house was a keen gardener, offering to keep the room well supplied with house plants. This has worked admirably and the instrument has caused no further problem.

The action is unusual, but not uncommon and is illustrated (see Figure 67) in order to warn of its particular problems.

The extended V-shaped hammer butt and cushion has two functions: firstly to act as a cushion for the jack, and secondly to cushion the blackleaded wooden slider which, when the hammer is raised, presses against the opposite side when the key is at its lowest point, holding the hammer against the check.

At rest as shown, the slider does not sit against the cushion as the jack does. The set-off is in the same position as in the 'D' type action and is adjusted in the same way. The steeply angled hammer rest comes half way up the hammer shank, well away from the head.

When the hammer checks, the jack is released from the notch, the slider presses against the opposite side of the butt until the key is released. When this is done, the tension of the spring raises the hammer enough to allow the jack to slip back under the notch again without the necessity of returning to the rest providing better repetition.

The jack is attached directly to the wippen (lever), has a spiral spring fitted to the tail as in an upright action and can be removed for repair or service in the same way as in the 'D' type action.

Fig. 69

The slider sits on a rocker carriage attached to the rear of the jack and is adjusted for tension from the keyboard side. The top screw increases the tension, the bottom screw reduces it.

Do not attempt to adjust the spring tension by bending it as it will snap quite easily. They can become brittle with age and although seemingly simple, they have a flattened end which is just pressed into the carriage and the wooden slider, which stops them twisting to left or right and is difficult to reproduce by hand.

Julius Blüthner Aliquot Patent grand Ser. No. 90478 *circa* 1913

This instrument deserves special mention on two counts: firstly the action, which causes some tuners difficulty chiefly because they don't know how it works, and secondly, how to tune the Aliquot stringing.

The Aliquot stringing extends from a point to the right of the central frame bar in the octave above middle C to the top A resonating in sympathy with the note being struck. It has its own damper and is designed to resonate only with its companion string.

With the exception of the last octave and a quarter, it has its own soundboard bridge and is tuned a stretched octave above its companion note, i.e. C5 (note 52) is accompanied by an Aliquot string tuned to C6 (note 64).

Where the Aliquot bridge finishes on the soundboard and resumes, by means of a metal stud, on the main bridge to top A, the tuning follows the same pattern as the main strings, i.e. A6 is

Fig. 70

accompanied by an Aliquot string tuned to A6, or almost, as will be shown.

The finest Blüthner specialist in our factory, Dave Brunning, was aware of my interest in obtaining the maximum quality for my work, and I feel that passing his word on is my way of thanking him for his patience with me.

To obtain the best effect from the last octave and a quarter of the Aliquot, the string is tuned the amount sharp that one would tune a stretched octave. This way, the slight edge of the note gives it a lift when resonating in sympathy,

When tuning the treble, first tune the basic note and pluck the Aliquot string with a finger nail, nothing metallic which will distort the sound.

The stretched octave will be discussed in detail in the chapter dealing with tuning.

Remember when restringing the Aliquot that the gauge of wire is not the same as its accompanying note, i.e. the Aliquot string of C5 is of the gauge with which C6 is strung, until the last octave and a quarter, when the Aliquot is strung with the same gauge of wire.

Blüthner Hopper action

With the treble action standard removed, the movement of the action can be seen. At rest, the right-angled hopper spring is

Fig. 71

Fig. 72

adjusted to hold the hammer at the correct distance from the strings and at the correct height above the jack to enable it to return under the notch of the hopper. Note that the hammers are not resting on the hammer rest rail.

The curved hopper spring does not at this stage touch the tip of the jack. The jack spring to the rear of the key sits in a slot in the felt pad glued to the tail of the jack. The other end is push-fitted into the soft wood of the key.

When the key is depressed, the jack sends the hammer towards the string until the point of set off is reached. When the hammer checks, the right-angled hopper spring holds the hammer against the check, assisted by the curved hopper spring.

As soon as the key is released, the tension of both springs raises the hammer slightly, allowing the jack to return under the notch of the hopper ready for repetition without the necessity for the hammer to return to rest.

There are other varieties of action and stringing, but this selection shows some of the more common varieties, which are met with and cause most problems.

6

PIANOS WITH PROBLEMS FROM NEW

It is quite erroneously expected that all new instruments will be fault free. This is often not the case and problems such as these are sometimes very difficult to solve.

Many instruments of cheap construction are purchased with an expectation of a far higher performance than they are capable of producing. There is no way of improving the performance of a poor instrument except on occasions when the design has been executed with a fault capable of correction.

In this chapter, we are going to look at a series of problems, some capable of a solution and some which are not. We will look at problems of bad design and ways of correction or minimising the effects of them. Never underestimate the irritation of even the smallest pianistic problem to the pianist. It is not the slightest use saying to the pianist that 'They are all like that'. If the instrument is incapable of producing the quality expected, then the reasons must be explained to the owner.

Alexander Herrmann upright Ser. No. 26601 *circa* **1970**

This instrument from the former DDR has many problems when put to the kind of regular use by many different pianists found in schools. The extremely sharp angle of the overstringing brings with it a considerable downward and sideways stress which with school use causes strings to break off at the agraffes, and constant replacements have to be made in both treble and bass.

Within two years of purchase for a school, due to the poor quality of the hammer felt and excessive use the pattern of hammer wear shown in Figure 73 was experienced which also served to increase the incidence of string breakages.

A close look at the wear pattern shows that the four hammers on the right have a pattern of indentation consistent with normal

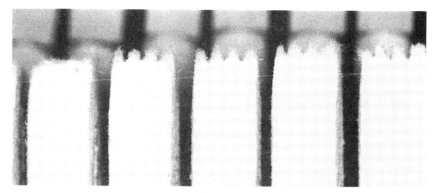

Fig. 73

Fig. 74

wear and tear, i.e. three deep cuts where the strings have bitten into the hammer heads.

If we look at the two hammer heads on the left, however, we can see a pattern of wear consistent with either loose flanges or loose centres, i.e. the heads showing no clear-cut grooves but a completely flattened head and considerably more wear due to the side play.

In both cases, as the wear is increased, the area of blow to the string is increased, which in turn is a cause of many string breakages.

The bass bridge, due to the sharp angle of the overstringing, has a right-angled section at the commencement of the central break, in order to place the bridge in a reasonable position on the soundboard.

The section commences with four sets of three steel strings, followed by two sets of three brass-covered strings and then two sets of three copper-covered strings, before the commencement of the bichord section.

Fig. 75

The shape of the bridge and the mixture of strings produces an unusual mixture of tone in this area and there is no way of altering this.

The fitting of a set of good quality hammers, together with recentering of the action, has reduced the breakages but not eliminated them.

In household use, the instrument provides reasonable quality which, for the purchase price, is competitive with the cheaper range of instruments in its range but does not have the quality or the durability for work beyond grade 6.

Challen upright Model 988 Ser. No. 102908 *circa* 1976
Height 40" (102 cm). Width 57" (147 cm). Depth $22^{1}/_{2}$" (57 cm)

This popular upright has in construction overcome some problems, only to create others; nothing however, which can't be overcome by knowing about them in advance.

The Kastner-Wehlau floating centre action and its associated plastic key bushes, has at least overcome the problem of excessive wear caused by 'dog-leg' keys.

The keys to the right of the break would, if bushed with felt, wear very quickly on the left-hand side, unless bushed with the plastic sprung bush, shown in Figure 75 removed.

In the centre section of the instrument, the stringing between the agraffes and the wrestpins can only be described as a jumble.

There are considerable problems for the tuner if constant checks are not conducted on notes already tuned to see if strings which are hanging together have been disturbed as its neighbour is tuned.

As the strings approach the central break, there are four bichord strings. Figure 77 shows that the wrestpins of the second pair from

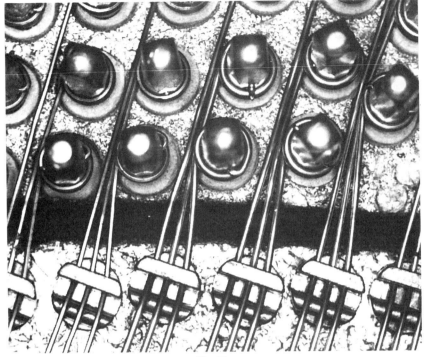

Fig. 76

Fig. 77

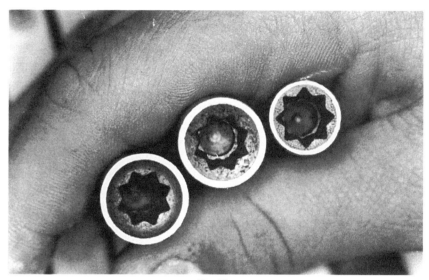

Fig. 78

the left are very close together, the upper pin being drilled lower than the ones on either side, too close in fact to take a standard tuning key head. On three instruments I tune, the top pin of the pair is loose. This however, may have been caused by previous tuners forcing a standard key onto this pin.

The three tuning heads shown are, from right to left, thin wall head, small head and standard head. Tuning the top pin, the thin wall head, will not disturb the lower string.

In general use, it is better to use a small tuning head, as this avoids the head gripping too far down the wrestpin and in the process touching a neighbouring string which has already been tuned. In this way, one of the self-inflicted problems of tuning is avoided.

The only problem to be guarded against in the use of the small head is when instruments have been fitted with larger wrestpins or when instruments have been subjected to damp and a greater grip is needed to move the pin, when the standard head should be used.

Too small a head can chew off the edges of the wrestpin, as can the use of old or worn tuning heads, which should be replaced before this occurs. Whilst not judging a book by its cover, a tuner can be judged by the quality of the tools used, and a tuner with cheap or worn out tools can only give the kind of attention to the instrument that these tools will allow.

Daewoo Royalle grand Model Rg-2 Ser. No. G018095 new 1988 Renner action

The owner of this grand is a pianist of international competition level. The problem which the agents had been unable to solve to her satisfaction was a 'zinging' sound in the centre of the keyboard when the key was released slowly.

On being asked to reproduce the sound, she played a piece which required a delicate touch and removal of the finger from the keys. It was during the removal of the finger that the sound came.

My first impression was that all of the dampers in the centre were of the W (split wedge) type, but a closer examination revealed that in the area which was giving the trouble there was a split wedge on the keyboard side, and a flat damper to the rear. The split wedge was exceptionally long and protruded through the strings much more than usual compared with other makes.

Fig. 79

Bechstein

Fig. 80

Steinway

The Bechstein and Steinway dampers do not protrude through the strings as much and the associated flat damper on the Bechstein is very much softer, expanding and contracting to the same level as the tip of the W-shape as the damper is raised and lowered.

Whilst being considerably older than the Daewoo, the associated flat damper on the Steinway still retains the downward curve from the stitched centre and is still almost level with the tip of the W-shape.

The flat damper on the Daewoo was nowhere near the tip and I felt that this was the cause of the zinging. As the damper was slowly lowered, the tip was hanging loosely between the strings before the flat damper caught up with it.

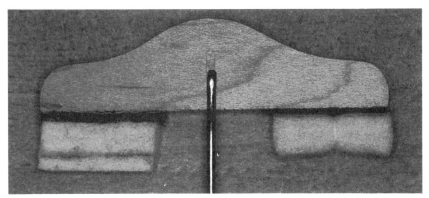

Fig. 81

As an experiment, I carefully removed the split wedge half of the felt from the note which was causing the problem, replacing it with a flat damper which matched the existing flat half of the damper, saving the split wedge in case this didn't cure it. The zinging sound disappeared, and as this seemed to do the trick and the section causing the problem was only short, I altered the remainder.

Where the split wedge was on both ends of the damper, the problem was not so pronounced and, although still present, caused the owner no problem. I advised her that flat dampers were not so efficient as split wedge and if at some future date she felt the need for more efficient damping, I would re-cover the section with split wedge which would not protrude so far through the strings.

To date, the dampers are working to her satisfaction. It took me two years to obtain string and instrument stability, together with the correct playing-to-tuning ratio. The pianist is a student and the instrument received surges of activity.

This problem of irregular use deserves a section all on its own and will be dealt with when we come to tuning problems. In this case, tuning is completed after the summer holidays in September, January after the Christmas holidays, and May after the Easter holidays. In this way, the tuning has time to equalise and is completely set by the time of the next surge.

The quality of tone produced, taking into account the price of the instrument, is excellent, and the pianist expressed complete satisfaction with the tone, which was what she chose it for in the first place.

In general, I do not like to redesign features such as this because it can cause other problems, but there are times when there is no alternative.

Kemble upright Ser. No. 227751 new 1988

This instrument was one of three supplied to Chester Pianos. When the first one was sold, they received a complaint that the notes were sticking. Of course, when their tuner arrived to test it, the notes would not stick but later the same complaint was received again. The notes would not stick for me either when I was asked to investigate.

In cases like this, the only way to pursue the problem is to ask the owner to duplicate *exactly* the situation which made the notes stick.

- When did the notes stick, all the time or occasionally?
- Exactly which piece of music was being played?
- Who was playing?
- Was there any heating on at the time?
- Were the pedals being used?

All or any combination of the above situations can provide different situations which can make the notes stick.

On this occasion, it was Sunday, the lady owner was playing and she was playing hymns. Although it was not Sunday, I asked her to play the hymn again to try to duplicate the situation to see if the same notes could be made to stick. From a cupboard, she produced a large and heavy hymn book and placed it upon the music desk, complaining at the same time of the lack of book holders. As she commenced to play, the notes began to miss occasionally.

Gently pressing down upon the desk, I could see that the hammers were leaving the rest rail and if I lifted the music desk, the hammers settled back further against the hammer rest rail. This, then, was the problem.

On this instrument, the top door, the front and rear fall, were separate, the fall receiving no support from the rigidity of the top door as it does in the type where the top door and the fall are one. In fact, the top door rests on the rear section of the fall and is quite heavy.

Fortunately, the remedy was simple and I fitted a $2^3/_4$" (7 cm) screw into the balance rail at the central break as shown in Figure 82. A small pad of bushing cloth was glued to the underside of the fall at the point of contact with the screw to mask any vibration, adjusting the screw until it was a firm fit. This stopped any sagging of the fall and pressure on the keys from the nameboard.

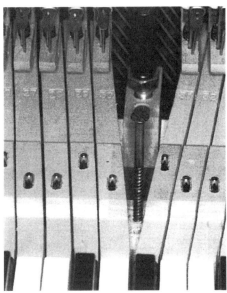

Fig. 82

This has caused no problem since. Being honourable dealers, I was asked to modify the instruments remaining in the showroom.

Bentley upright Ser. No. 106269 *circa* **1964 Richard check action**

Hammer butts and balance hammers should be re-covered in doeskin, or whatever skin passes for the re-covering in action leather today. Most students are shown the correct way around to place the skin to the wood.

Manufacturers of the cheaper range of instruments and some repairers resort to the use of mock doeskin to reduce costs. This material is much the same as cotton moleskin in the manufacture of men's trousers, it is easier to fit, as well as being cheaper to purchase. In use however, the material rubs off, leaving a brown residue which showers over the jack slap rail and set-off rail and is dusted away by tuners without noticing it.

The material will last in normal household use for about twenty years. The same instrument in conditions of excessive use such as schools or clubs, can show considerable signs of wear in a very short time, six to eight years. In the same conditions of use, real doeskin would last up to sixty years, more if in moderate use, as is seen on many older models.

Leather with a hard outer and a soft inner is sometimes substituted for doeskin. It will last longer than mock doeskin but can cause problems if not correctly fitted.

Fig. 83

The owner of the above instrument had already had it tuned to his satisfaction but felt that the touch left a lot be desired. The previous tuner could find nothing wrong and had completed the tuning and regulation in a most satisfactory manner.

A quick check confirmed this, but when I started to play, there was indeed something wrong with the touch; shock waves travelled back from the action through the keys to my finger tips. In swift repetition, the jack jolted on the notch as if there was no covering on it at all.

Removing a central hammer, I examined the notch and found that it had been covered in leather which had a hard skin side and a soft inner. But instead of being fitted hard side to the wood, soft side outwards, this was fitted with the hard side to the jack, and it was this which was causing the hard touch.

Beneath the leather covering was a very small and insignificant cushion intended to protect the point of impact of the jack upon the butt, another reason for a harder touch.

No attempt should be made to improve upon the cushion as, combined with the increased thickness of a new doeskin covering, there is no room for a great deal of regulation to take this into account.

As a demonstration of the difference that a real doeskin covering would make, the leather was removed with a Swan-Morton No. 11 scalpel, taking care to clear the slot above the cushion felt. This

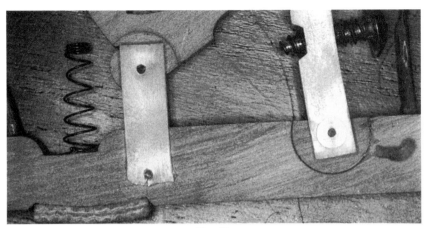

Fig. 84

slot holds the replacement doeskin in position as it bends around the butt.

When the butt was re-covered, it was replaced, the capstan, set-off and check were regulated to the new covering thickness and the touch was completely different by comparison with the notes on either side.

As this was a manufacturing fault but the instrument was well out of guarantee, the owner decided to pay for the re-covering of the butts.

The total time spent on disassembly, recovering and reassembly, was six hours, charged at workshop rate, with a further two hours for regulating and tuning charged at standard tuning rate.

Flanges are another source of problems on this kind of action. If they are viewed from the side, it will be seen that they are made from a soft plastic and have no bushings. This soft plastic needs no bushing or lubricant. There is no bushing in the hammer butt either as in some older instruments, the centrepin is held tightly in the butt and moves smoothly within the flange and wear is minimal. Although this may seem ideal, great care must be exercised in spacing hammers or replacing parts to avoid over-tightening the screws.

These soft plastic flanges can split at the screw hole. A little over finger tight is all that is required. The chipboard self-tapping type screws are probably the best part of the action as they hardly ever come loose in use. It is the success of this type of screw which prompted me to use them as replacement screws on D-type grand actions.

The H-type jack flange on the wippen is also made of plastic, and cannot be glued to the body. It is riveted to the wippen. Many of them have been split in manufacture but at the moment are causing no problem so I have left 'well enough alone'. If they cause trouble at a later date, I will replace them with a new wippen section complete with jack which fortunately are available as replacement parts. (*Note*: the existing jack will not fit on to the new wippen, the flange is of a different size.)

The pressure which has split the flanges at the factory is the pressure which must be avoided when recentering during repair. The hand-held centrepin pliers with the shaped pin are useless for this purpose, as they tend to split or break the flange. The bench type decentering machine with the long thin needle will avoid damage.

These actions when used in schools have necessitated the complete replacement of all hammer flanges and their replacement with wooden ones. The big problem in doing this is that due to the need for thick sides on the plastic flanges, the hammer butt has been made thinner and the nearest replacement flanges have a larger central gap, which allows the butt to move slightly sideways within the flange.

In individual cases, I have successfully packed the gap with punchings of silicone paper from my computer labels made with my small desk paper punch, but this is a tedious, time-consuming and expensive job.

The hard nylon based flanges which contain normal bushings seem to cause no problems we do not already encounter with wooden flanges, but are not a replacement for the soft plastic type.

The only other constantly recurring problem with this model is with the pedals. The small bottom door stands directly on the bottom pedal board with no vertical aid to support it other than the two notches which stop it falling forwards. With constant use of the pedal, the panel begins to slide up and down as the pedal board loosens, requiring constant waxing of the notches, side grooves and top support and door catches.

Wilhelm Steinman Ser. No. 359514 new 1987

The reason for my call to this instrument was a strange knocking coming from the keys in the centre of the keyboard.

This medium priced upright from the former DDR has the standard size of frame of many other instruments, and on removing the top and bottom doors revealed a well-designed frame and action. The most impressive thing which caught my eye was the precision of the wrestpin stagger and the excellent string spacing. It was refreshing to see that there was not one string touching its

Fig. 85 Fig. 86

neighbour, all the coils were at the same height and there was no need for agraffes to keep the strings separated.

In the period before unification, these instruments were heavily subsidised by the East German government both in price and labour in order to gain access to hard currencies. In creating jobs, it is possible that more time was spent on these desirable items than in our more cost-conscious Western methods of production.

There was an obvious knocking coming from the keys in the centre, caused by the sharps tapping against the nameboard due to a warping in the centre of the rail. The method of correction has been borrowed from Bentley uprights of the 1960s.

A metal eye was fixed to the rear of the nameboard at the central break, a $2^{1}/_{2}$" (6.5 cm) screw was pushed through a thin front touch washer, inserted through the metal eye and secured directly into the balance rail. If this is done at a slight angle, the screw can be tightened just enough to keep the rail away from the sharps. Do not screw it down too tightly, as this will press the nameboard down against the top of the keys causing them to stick. The felt washer will stop any vibration of the screw and the metal eye.

My wife was always accusing me of having pockets like a little boy, full of rubbish. My tool box is full of what other people think of as rubbish and throw away, but by keeping such items to hand I am never at a loss to find just the answer to a small problem such as this.

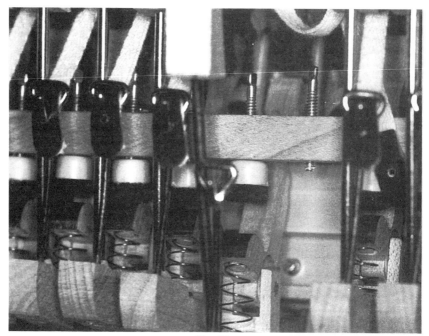
Fig. 87

Daewoo upright new 1990

Direct import sales of pianos such as Daewoo and Offenbach means that by the time the instruments reach the dealers, they are howlingly out of tune and the dealers are faced with having to have them tuned before they can hear the quality of sound produced.

Having been called to a local dealer to tune the above instrument, I found that some of the notes did not work. On opening it up, I discovered that one of the set-off buttons had fallen off, and was impeding the movement of its neighbour.

A closer inspection showed that many of the set-off buttons were hard up against the set-off rail and an attempt had been made – probably at the factory – to adjust the set-off higher than the button could raise, eventually turning the adjusting screw out of the button. The height of the adjuster can be seen in Figure 87 as can the rest of the buttons close by. The set off button was replaced, but it would not adjust high enough to set the hammer off at the correct distance from the strings, without modification.

The hammer was left as close as it would go and the tuning was commenced. As I reached the wound strings, I found myself attempting to adjust the wrong string. In cases, like this, I accuse myself of carelessness. When I found myself on the wrong string again, it bothered me but when it happened a third time, I stopped tuning altogether. This was most disconcerting.

At first, I could see no reason for this stupidity of mine and spent some time staring at the bass strings, wondering what on earth was wrong. Suddenly, it dawned on me: the wrestpins in the bass commenced with the top pin not the bottom pin and it gave the impression that the strings were in pairs as usual, but in fact the seeming pairs were strings belonging to separate notes.

Daewoo and Offenbach stringing Fig. 88

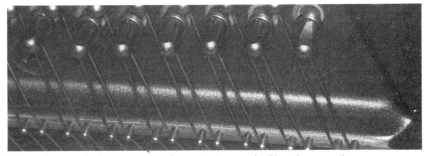

Yamaha and the standard stringing of all other makers Fig. 89

The tuning was continued by following every string up to its appropriate wrestpin. My first thoughts were that this would confuse the inexperienced tuner, but as I progressed I realised that I was doing just what they would do anyway and that it was I who was getting confused.

After fifty years of wrestpin formation in a standard pattern, I was suddenly faced with a complete reversal. Only once before have I experienced this reversal. That was on a Canadian Bell miniature piano and I was not impressed with that one either.

EXPERIENCED TUNER, BEWARE!

Fig. 90

Starr upright new 1987

This instrument of Chinese origin glorying in the name of Starr was given as a prize in a piano competition by a local music house. It was causing the pianist considerable problems, all caused by bad design and cheap construction. The owner felt that under the circumstances it was not possible to complain, and as the tuners sent by the music house had singularly failed to solve the problems, I was asked to investigate.

A glance inside showed hammers which were badly spaced and wippens in much the same condition. The hammers rested irregularly against the rest rail and this problem was tackled first. Having spaced the hammers and adjusted for the uneven resting of the hammers against the rest rail, it was in the spacing of the wippens that the cause of the problem was revealed.

Although a metal flange rail (beam) is a good innovation in action construction, it is only of use when the flanges fit correctly. It was obvious here that they neither fitted flat to the rear of the rail nor did they fit tightly in the flange groove. There was no way that this could be corrected, without packing the flange rail with box cloth, an expense that the owner could not afford.

Some of the jacks had cracked due to being used at an angle to the vertical and had caught on an adjoining note, the centrepins

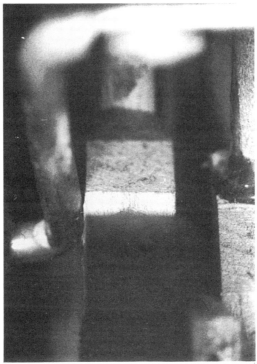
Fig. 91

were hanging out to one side. New jacks were fitted and the action screws screwed down as tightly as possible.

The half blow mechanism on this model moved the whole rest rail forward and was held in place by three cranked brackets. The one in the centre however, was causing a problem. As can be seen in Figure 91, the angle of the bracket was too acute and was catching on the hammer butt. In fact it was possible to move the whole rest rail from side to side.

In order to remove the bracket, the hammer had to be removed from the action and it was at this stage that the mark made by the bracket on the side of the butt could be seen.

After recranking the centre bracket to a better angle and also checking the brackets at either end to stop any side play, it was refitted together with the hammer and has caused no problem since.

Every time this instrument is tuned, the hammers and wippens have to be respaced, but this could not be avoided without major modification.

Instruments of this type (many look as if they have been made by MFI and some sound like it) are imported due to their low cost and bought by unsuspecting purchasers for the same reason, but it is the tuner who has to correct the problems and satisfy the complaints of the pianist.

In this small selection of new-piano problems, we have looked at silly little difficulties which should have been sorted out at the design stage. We have looked at pianistic problems which perhaps could not have been foreseen and we have looked at the problems of bad and cheap design. It is a pity that there seems to be no way within the trade for these things to be fed back to the manufacturers.

7

HUMIDITY

This problem, which is so often overlooked when fault finding, is the cause of a considerable number of customer complaints yet is so easy to remedy. Humidity is essential in some form for most instruments, even if it is just some potted plants in the same room to relieve a dry atmosphere. It helps to stabilize the tuning and condition the soundboard to prevent cracking.

We will look at instrument design, the effects upon it of central heating and the means whereby the resulting damage can be avoided or minimised. We will study individual case histories and how humidifying has relieved a troublesome situation.

If we examine some of the wrestpin formations at various points in both new and old instruments, some of the reasons for split wrestplanks and loose wrestpins will emerge.

Sames upright *circa* **1910**

A glance at the wrestpin formation shows that they follow the line of the grain of the central wrestplank lamination, i.e. horizontally. No attempt has been made to stagger them, consequently there are three lines of holes drilled along the grain of the second lamination with very little space in between.

Fortunately, there was a random pattern of loose pins, not a single line, showing that although the central heating had taken its toll, the wrestplank had not split.

In its present state, the tuning would not last more than a few months and it was decided to humidify it to stabilise the pins and wrestplank.

Although there are expensive humidifiers on the market, it is often sufficient, if there is room, to place a 1–litre plastic ice-cream container filled with water in the bottom near the pedals. This will supply the needs of the dry timber if the instrument is kept closed except when in use. The rate of absorption can be up to

Fig. 92

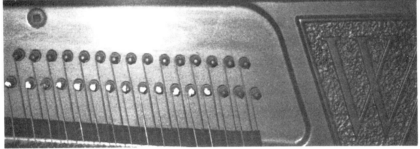

Fig. 93

three pints over a four-month period, so the container must be examined at least once a month.

It was decided to leave the instrument to humidify for at least three months. At the end of this period, the tuning was sufficiently stabilized to last for four months and as time passed, the tuning period was extended to six months without problems.

Wrestpins fitted without stagger are bound to weaken the timber of the wrestplank between the pins. This is a problem experienced in some degree by many makers and is most prevalent in the region approaching the central break.

Bass section of Waddington upright *circa* 1920

The wrestpins marked here with chalk, show that the wrestplank has cracked along the central lamination in line with the lower wrestpins. There are many Waddington and Bremar uprights affected in this way and I always test the pins in this area before attempting a tuning which may be impossible, or which may indicate need of major repair.

In this case, because of the quality and condition of the remainder of the instrument, the owner was warned that there was still a danger of the wrestplank splitting further in spite of my

Fig. 94

ministrations, but with her concurrence larger and longer pins were fitted in an attempt to bridge the gap. This has lasted for five years to date, and with the humidity kept up, has caused no problems.

Monington and Weston upright *circa* 1930
Treble break

Trouble caused by the close fitting of wrestpins vertically is very often a feature of older straight (vertically) strung uprights and is a point to watch for loose pins. The timber weakens in a vertical line and without the support of the substantial Monington and Weston frame, causes problems.

Bechstein upright Ser. No. 55671 *circa* 1900

Bechstein uprights are not immune to problems in this respect. It is only by humidifying the instrument illustrated in Figure 95 that it will hold its tune. It can be seen that the problem areas lie at either side of the central break. If the problem exists these are the places to look and the places to inspect before tuning commences.

Fig. 95

Steinway grand Model B Ser. No. 424852 *circa* **1971**

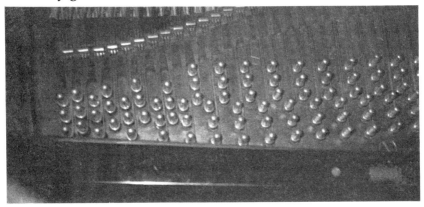

Fig. 96

The tremendous trouble that Steinway take to stagger the fitting of wrestpins in this model until they reach the central break shows the importance that they attach to the problem, but even they have difficulty as they approach the central break.

This instrument is kept in an art gallery, and although there is no difficulty with the wrestpins, pianists had been complaining that after only twenty minutes playing the instrument was beginning to go out of tune. Suspecting that the gallery heating was the cause, I decided the only way to stabilize the tuning was to humidify it.

The humidifying of a grand piano is not as simple as an upright, there is no room for the water inside and the room was far too large to humidify as a whole. It was decided to cover the whole instrument down to the floor and fit a hanging humidifier suspended beneath the soundboard or as in this case, place a large tray of water on the floor underneath.

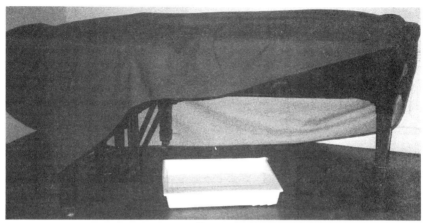

Fig. 97

In spite of the fact that the instrument is removed from venue to venue for concerts, it retains its stability and has caused no complaints for the past fifteen years since humidification.

Yamaha upright Model U1A new 1986

This excellent instrument was difficult to settle down, but not, I hasten to add, due to any problem of construction or quality.

The surges of playing it was getting during the Christmas, Easter and midsummer holidays was difficult to relate to the playing-to-tuning ratio. In the end, tuning was related to these surges, being completed immediately after them, allowing time for the tuning to stabilize before the next surge.

This however, did not supply the complete answer. The pattern in which the tuning deteriorated was not consistent with playing alone; the tuning deteriorated more in the winter than in the summer. Although I was assured that a high level of heating was never kept in that room, the pattern seemed to indicate fluctuation of temperature and a need for humidity.

The owner was a young lady, and the room small. Like so many young ladies, she preferred to turn up the heating when the weather turned cold instead of adding a layer of warm clothing. The house was unoccupied during the day and a high level of heat applied during the evenings. I felt that it was this constant changing of temperature that was affecting the instrument so badly. My advice was to leave a low level of heat on during the day, which in turn would not call for such a high level of heating during the evening, and cost of which would even itself out over a period of 24 hours.

Humidity was commenced with a 1–litre container of water

placed in the space to the right of the pedals. Due to the amount of use it was getting, the instrument required frequent regulating until the felts packed down. Once this had been achieved, there was only the level of humidity to be considered. Gradually, the humidity started to take hold and the tuning began to stabilize. A larger container was substituted and the tuning stabilized.

Much of the humidifying depends on how reliable the owner is in keeping a watch on the water level. This must be kept up all the year round, not just in the winter, and the instrument must be kept closed when not in use to keep the atmosphere constant.

The tuning of this instrument varies very little since humidifying, in spite of the surges in use, the playing-to-tuning ration has been established at three tunings per year and although it is just starting to show signs of deterioration as the tuning becomes due, the owner is not complaining, recognising that the amount of playing justifies it.

Bechstein upright Ser. No. 13672 *circa* 1882

Height 52" (132 cm). Width 61" (160 cm). Depth $25^1/_2$" (65 cm) Reconditioning of instruments of this age is fraught with danger. After eighty years or so, according to quality of make and construction, there is a constant danger of structural faults developing, which may involve a further expensive repair or even damage which it is not possible to repair effectively.

The instrument in Figure 98 had been expertly reconditioned and the problem complained of was just that there were a few notes sticking occasionally and that a soft 'woolly' sound had developed in the bass. In this model, the action is tied to the key carriage by a slot in the base of the prolonge (sticker), and the action cannot be removed completely without first disconnecting the prolonges from the key carriages.

Access to all parts can however be obtained by releasing the action catches and pulling the action forward. There is enough room to reach the damper and wippen screws whilst still remaining attached to the keys.

Having detached a single key which was sticking, the problem was revealed, namely, corroded key weights. Filing the corrosion away eased all of the sticking notes.

A test of the bass revealed a decidedly 'woolly' sound which commenced immediately past the central frame bar. It is at this point that the bridge is shelved back to a better position on the soundboard. It is an exceptionally large apron and this, coupled

HUMIDITY

Fig. 98

Fig. 99

Fig. 100

with the age of the instrument and the effects of central heating had caused the bridge to sink.

A crack had developed along the line of the apron and the joint

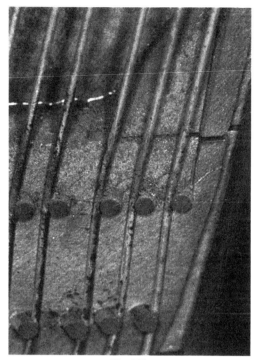

Fig. 101

connecting the long bridge to the apron bridge had separated. To effect a repair, the strings would have to be removed, the apron and bridge repaired and the strings replaced. This type of repair would be expensive and on top of the initial cost of purchase or reconditioning would have to be paid for by either the vendor/reconditioner or the repairer according to a particular situation. In this case, it was out of the guarantee period and the owner had to pay. Even after such a repair, no guarantee can be given that other structural faults will not develop.

Look closely at the bridge on all vertically (straight) strung Bechsteins of this age. If a crack seems to be developing, warn the owner and humidify the instrument to stop it drying out.

Bechstein upright Ser. No. 78857 *circa* **1905**
Height 52" (132 cm). Width 61" (160 cm). Depth $25^1/_2$" (65 cm)

This later model did not have the problems of the last, earlier model due to the strengthening of the bass bridge apron. The two extensions to the apron give considerable support and avoid the problem of the sinking bridge experienced in the previous model.

However, the vulnerability of the junction between the long bridge and the apron remains. At the first tuning, an inspection

Fig. 102

of this point showed that a very slight crack seemed to be developing, and as a precaution the owner was advised that the instrument must be kept humidified in order to keep the effects of central heating at bay and the glue at the correct level of humidity.

This served to keep things going for five years until a cold snap in the winter of 1990–91 caused everyone to raise the temperature in their households considerably. On receiving a call in April 1991 to the effect that the piano was making a funny sound and could I bring the tuning forward? My immediate reaction was to ask if the humidity had been kept up. The reply was, 'No, I've forgotten, but I've put some in now.' If the humidity had been kept up, there would have been no problem.

My first inspection was of the bridge joint, finding it completely shattered. The joint had opened up, the bridge pins and top third of the joint had moved to the left due to the pull of the strings, and two of the notes to the left of the joint had also weakened. The point at which the strings have lost contact with the long bridge can be seen now that the strings have moved to the left. Although the bass bridge has not sunk, it is still a major repair.

Whilst not advising the repair of an instrument of this age, if it is decided upon, it will be difficult to replace the top section of the joint. The bridge pins will have to be removed, a hardwood cap will have to be made to the previous pattern for the last four notes, redrilling for bridge pins and restringing.

This is yet another case of the age of an instrument being against it in our modern conditions of central heating if humidity is ignored.

In advising customers who are considering the purchase of an instrument of this age and type, inspection of this vulnerable point as well as the usual items of wrestpins, soundboard, wear and

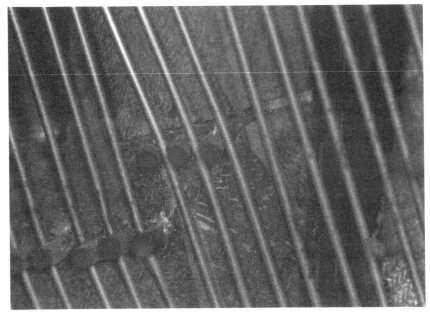
Fig. 103

tear, etc. must be made and warnings given of the problems of keeping such an instrument in conditions of central heating.

For pianists of any standard above grade 4, most vertically strung instruments lack the quality and range of sound to produce the shades of tone between treble pianissimo (*ppp*) and treble forte (*fff*).

Although humidity will be dealt with again under the heading of Problems of Tuning, when we deal with string and instrument stability, this chapter has been designed to show the successes when humidity is taken into account and the failures when it is ignored.

8

TUNING: ADDING THE PROFESSIONAL TOUCH

Many tuner/technicians leave college with a very heavy hand, inflicting upon themselves untold problems of their own creation, adding to the already considerable number inherent in all instruments. Many are completely unaware of the fine-tuning tests available for the smoothing of the scale or bearing, being content that they already know enough to 'get by'. Some methods of tuning being taught leave a lot to be desired, turning out what used to be termed 'rough' tuners, depending on the music houses to supervise and continue the training started at college.

This will be the largest chapter in the book, simply because it deals with the greater part of the tuner's work and in general from my research requires the most effort in post-college training.

The multiplicity of headings will not be enumerated here and I do not ask that tuners change their methods, but if their work can match the tests of scale and octave we will study later, then they have no need to worry. If not, then there is a need for further study and practice.

Self-inflicted problems

We start our tuning with the setting of the scale or bearing, and it is in this area that I have seen tuners perform the most diabolically destructive operations on the strings in preparing to tune an instrument.

The use of a muting strip is perhaps allowable in the learning stage, although I have never allowed my apprentices to use this method, as it distorts the first and third strings of the notes in the bearing to the extent that any vestige of tuning remaining is immediately destroyed. You only have to hear one being ripped out to cringe at the thought of the damage being done.

The preservation of tuning and instrument stability go hand in hand. Every item of tuning preserved is one unnecessary string movement avoided. On an instrument which is regularly tuned, it is inconceivable that every string will go out of tune every time.

If the correct tuning-to-playing ratio is maintained, and attention paid to humidity, string stability will bring with it instrument stability. The experienced and professional tuner knows this and it is the reason for the warning I received from my senior tuners not to 'Pull them about too much', so think CORRECTION and PRESERVATION.

To raise the pitch of an instrument by just one hertz (cycle) is enough to destabilize an instrument and, unless necessary due to lack of tuning, should be avoided.

To determine the condition of the bearing, check to see if there are any strings of C or A which will consonance with the tuning fork, and start from there. Complete the unisons and carry on to the next note in the cycle of 4ths and 5ths. If any string provides a good interval, complete the unisons and carry on in this manner. You will be surprised just how many strings are just as you left them last time.

If the original starting point is correct, tuning time is saved and both string and instrument stability is preserved. If, however, instrument stability has not yet been achieved, then it is possible that all of the strings in the bearing area will have to be tuned and it is only when the extreme bass and treble is reached that strings will be found to be in tune.

The accuracy with which the original string is consonanced to the tuning fork is critical. If this is done with great accuracy each time, stability will come more quickly.

If tuners insist on striking the tuning fork on a hard surface, it is small wonder that the fork soon becomes out of tune. By striking it on the flat of the knee (some tuners use the heel of their shoe), the fork receives enough impetus to sound for quite a long time.

Extreme accuracy can be obtained by holding the vibrating fork between one's teeth and lower lip, the vibrations travelling around the bones of the skull, allowing both hands free, one to strike the note and the other to turn the tuning key. This needs some practice but the increased accuracy is well worth the effort.

When we come to consider tools for the job, we will see how we can further avoid distortion of the strings which can come from using even the best equipment.

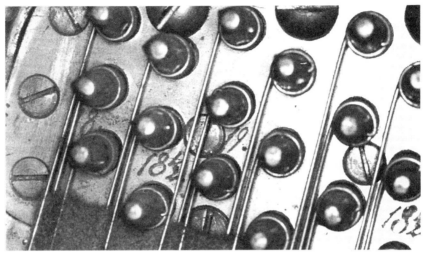
Fig. 104

Design problems within the bearing

Here, we are considering the tuning of the bearing or scale and any problems encountered. According to design, we may or may not find modifications to strings by way of length, thickness or overwinding within the bearing area, all of which tend to distort the very partials to which we are tuning.

This distortion may appear in the form of apparently correct bearing intervals which fail to extend correctly downward or upward from the bearing. In cases like this, a slight adjustment of the modified strings to correct difficult intervals above or below the bearing will often effect the compromise necessary due to this design problem.

The easiest pianos to tune are the ones whose bearing falls within the range of unmodified strings.

If we look at the stringing changes of the above Blüthner upright as it approaches the central break, it can be seen that if any of the strings in this area fall within the bearing, they must be suspect strings for further adjustment if the bearing fails to extend outwards from the centre successfully.

By testing the accuracy of the 4th and 5th intervals within the bearing, by use of the 3rd/6th and 10th interval progressions upward and downward from the bearing, any error will show up.

Remember that a tiny change in a 4th or 5th interval will make a tremendous difference in a 3rd/6th or 10th interval. No two intervals react in the same way to a given change in tension. If a wound string is within the bearing area, then it will most certainly be suspect.

Fig. 105

A close look at the Kimball piano (see Figure 105) shows that within the area of the bearing, the first six pairs are wound, the next four are steel bichord, and two trichord.

The owner was having great problems in obtaining good tuning, and I must agree that it was not easy to smooth out the bearing and obtain a successful progression outwards from the centre, but after three or four tunings, the best compromise was obtained.

The over-stressing of these instruments does not help with the tuning problem and although it is possible to obtain fairly good tuning, there is a considerable lack of quality in the overall tone, and a harshness that is often criticised by pianists accustomed to the European, more rounded tone.

Bass tuning problems

Extending the bearing towards the bass comes up against the problem of windings which have oxidised due to damp. From the engineering point of view, it is bad practice to place two dissimilar metals together in this way for just this reason, but from the point of view of tone and flexibility this is the only way that these requirements can be produced and we have to put up with the problem, avoiding as much as possible the cause, DAMP.

The problem manifests itself in the form of a dull tubby sound instead of a resonant boom, and is more frequently found in the double wound strings of the lower bass.

Windings which are loose or buzz can sometimes be improved by slackening off the string, adding a half turn in the direction of the winding and retuning. If this is not successful, then replacement is the only alternative. If this occurs in the bichord section, then it is advisable to change both strings, replacing with two wound on the same machine in order to ensure that the windings agree.

It is this agreement of windings, or lack of it, which brings us to the next frequently occurring problem. From the tuning point of view, the effect is that although the first string of the bichord can be tuned and all the intervals made correctly, when the second string is brought into unison, all the intervals are thrown out. If the second string is then tuned correctly with the intervals, the resultant unison is badly out. This is a sure sign that it is the windings which do not agree and both strings need to be changed, once again being replaced with two wound on the same machine.

One word of warning here: unless the customer complains, make the best compromise and leave it at that. Although it may be a problem to you, by bringing it to their notice and making the necessary changes at a charge which will allow the second visit to refit them and another visit to retune at some later date, you may be accused of correcting a problem which was causing them no bother and ending up with an out-of-tune piano which does cause a problem. This has happened to me.

In replacing bass strings, the new string will have a far better tone than the old one, and can have the effect of making the other strings in the area sound worse than they seemed when they were all equally poor.

Double-wound strings used singly in the bass have a winding which must agree with the intended frequency. If it does not agree, when the string is struck, there can sometimes be heard a particularly strong vibration which will be the reason for the failure to produce good tuning in the bass.

What happens is that just when the note is coming in to tune, another vibration starts up just as strong as the one being tuned. The problem then is to decide which is the correct one. Sounding the double octave helps, as does deciding which vibration moves more quickly with the least tuning key movement. TUNE TO THE SLOWEST-MOVING VIBRATION.

Once again, if this is a problem only to you, make the best compromise and leave it at that. If this is a customer problem, try slackening off the string and giving it a half turn in the direction of the winding. If this fails, replace the string.

IN ALL CASES OF THIS KIND, make doubly sure that the loop of the string is flat against the bass of the hitch pin before raising it back into tune.

Do not attempt this operation on old or badly rusted strings without warning the owner of the serious risk of breakages and the cost involved.

Where strings are rusted, release the string just enough to break the rust bond between the string and the metal of the frame before raising up to pitch. The 'chink' that can be heard as this happens is a sure sign that without this precaution, the string would be subjected to strain between the bearing bridge and the wrestpin, due to the rust bond.

Treble tuning problems

One of the problems most frequently met with in the treble appears in the region of the treble frame bar and the point at which the dampers end, sometimes higher. It manifests itself in the form of a spurious vibration of a few hertz (cycles) which will not tune, it cannot be altered faster or slower and can appear on a single string or all three. It is not particular to one note or harmonic series, and it can disappear and reappear again at a higher frequency.

The main feature of this phenomenon is that nothing will alter the frequency, damp it down, or increase it. Damping off all the remaining strings on the piano does not diminish it in any way. It is not sympathetic; if it were, it would not appear at the same frequency on so many different notes and would come and go with an alteration of the tuning.

It is not caused by the close proximity of the frame bar, as it is present on the excellent Knight upright which has no frame bar in the treble. It is not in any way connected with the iron frame (plate) as it is also present on wooden framed pianos. It is not the property of any single string, as replacing one makes no difference, therefore, it is not inharmonicity (the inability of a string to produce a particular partial correctly).

This phenomenon seen on an oscilloscope, has the same form as an echo and this would seem to be in keeping with the swelling and dying of the sound intensity as it decays.

Many methods of disguising this sound have been proposed, such as tuning one string of the three to try and counterbalance the sound with another set of vibrations of the same frequency. Over a period of time I have tried this method, and have only succeeded in muddying the tone.

Pianos of all qualities suffer from this problem, and always occurs in the same place. It may be produced by the distance that the strings are placed above the soundboard at this point. Possibly a baffle between the strings and soundboard would improve matters, and this is a thought for some experimentation.

Whilst not offering any solution, the problem can be minimised by tuning each string separately, starting with any string of the three which does not cause this vibration, finishing with the troublesome one or two to get the best possible compromise. Leave the note beating at the frequency of the spurious vibration which cannot be altered.

The only other recurring problem seems to be obtaining good unisons in the last octave and a half of the treble. Times without number, following other tuners, I find that the first string of the three is tuned very well indeed, but the other two are very badly out of tune. The piano did not get this way by accident or use, it had to be put into this condition.

If it is possible to tune one string well, it is possible to tune every string to the best the instrument is capable of. The confidence in one's ability to tune unisons well is often ill founded and I frequently tune the three strings separately as I would tune the first one in order to ensure the best consistent results, then checking the unison.

The tuning wedge

Probably the most varied and discussed item of the tuner's kit is the tuning wedge, each tuner having their own preference, often designing their own. When considering items such as this, I can only show my own favourites and explain the reasons why they have become so.

As we finished the last section with the need to tune each string separately to obtain good unisons in the upper register, let us start by showing how this can be done with a Papps wedge.

The strings at this point get close together near the bearing bridge, but are much wider apart lower down. The wedge fits more easily and does not disturb the strings.

To tune the middle string, the wedge can be splayed across the two outer strings quite easily as shown in Figure 107.

If each string in the upper treble is tuned in this fashion, placing less reliance on one's ability to hear unisons, an improvement in the overall quality of the tuning will result. Smoothing the unisons from this point is relatively easy as they are very close to start with.

Whilst on the subject of the Papps wedge, let us consider how it comes from the supplier. In its new form, it is sometimes impossible to push it in between strings without forcing them apart and as we know, we must avoid this at all costs. Over a period of years, I have tried to reduce the tip of the wedge to workable proportions but nothing matches the natural wear which comes from regular use.

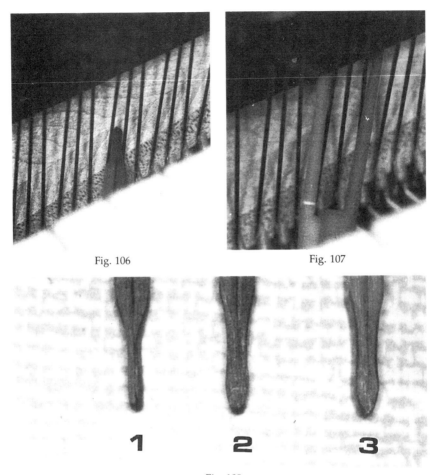

Fig. 106 Fig. 107

Fig. 108

In my case, I carry three wedges numbered 1, 2 and 3. My No. 1 wedge is reserved for instruments in which I have been able to produce the maximum stability. No. 2 is my general use wedge for use on instruments of lesser stability, not necessarily of lesser quality but whose stability has some way to go. No. 3 is or was a new one, reserved for rusted strings, strings of irregular spacing, first- or second-time tunings when the disturbance of the strings is not critical. Figure 108 shows how they get worn down to manageable proportions.

Where strings are badly spaced due to poor wrestplank pinning, it is unwise to respace them if the hammer felt is cut by the strings as this brings the string into a new position on the hammer, which will place an extra strain on the string and give a much softer sound. It would be wise to leave well enough alone. In cases like this, I use only one tine of the wedge instead of two, disturbing the strings much less. To use my No. 1 wedge for this purpose

Fig. 109

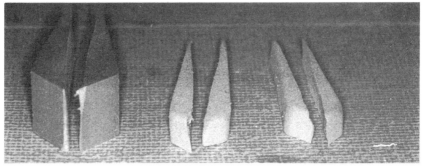

Fig. 110

might break it, and I would not risk that as this is my most prized possession.

On grand pianos, and in the bass of large uprights, a softer wedge is required which will absorb as well as dampen the sound. Felt wedges which are capable of being shaped to the desired length and thickness are usually too hard to absorb as well a damp and have to be forced too tightly between the strings to be effective. The felt soon forms ridges which pluck the strings when being removed, destroying the tuning. Rubber wedges as supplied by piano supply houses all seem to be too hard, and need to be wedged tightly in the strings to be effective. A softer rubber wedge which will provide damping and absorption by its own weight alone is required. The rubber must be soft enough to absorb the sound but not soft enough to crumble in use and heavy enough to provide damping by its own weight alone.

A selection of wedges is shown in Figure 109. Top left are two wedges supplied from two different piano parts suppliers; both are too hard. Top right are two experimental spring wedges made from a tuning gang. Bottom left are two grand wedges made from a pencil rubber purchased from Woolworths, and bottom right, four upright wedges made from a pencil rubber from W.H. Smith.

Some tuners make a long slender rubber wedge which can pass between the strings, resting on the soundboard, the angle of the

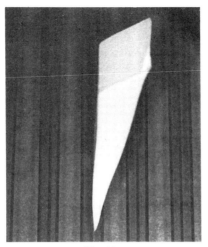

Fig. 111

wedge against the strings rather than the pressure being enough to provide damping without disturbing the strings. My own preference is for a wedge which by its own weight alone can dampen and absorb the unwanted string sound. The larger wedge as shown in Figure 111 can be *placed* in position, and will stand upright between the strings without forcing it down, its own weight being sufficient to do the job. If it has to be pulled out, it has been positioned too tightly or it is too hard.

The lighter, thinner type of wedge can be used on upright pianos and will not fall out due to its light weight. It will also flex behind the strings without distorting them.

Fig. 112

To aid in the production of better unisons in the treble of grand pianos, spring wedges made from a tuning gang are most useful.

The white portion of the spring has been cut from a soft pencil rubber which enables the tines of the wedge to spring outwards using this pressure to dampen the strings. They can be fitted without undue pressure and removed in the same way.

Fig. 113

Sympathetic vibration

Invariably, at sometime or other every piano suffers from some form of sympathetic vibration, not necessarily emanating from the piano itself, but being generated by the sound waves as they strike another object which is in some way loose or of the same wavelength.

Among common causes are brass hinges with a steel core, which, due to the expansion and contraction with changes of temperature, become loose. These can usually be cured by tightening the hinge screws or punching the central steel core with a centre punch into a new position.

Early Eavestaff minipianos had chrome lights fitted to either side with candle type bulbs which made a dreadful noise and were impossible to cure.

The sides of these instruments are hollow, and as the glue which has run down inside dries out, there is a tendency for it to buzz. This can be felt if fingers are pressed lightly to the side from which the sound comes, but there is no cure as there is no way of getting inside.

One extremely powerful Steinway upright shook eveything in the room, including the glass door of a cabinet right next to it. The door was secured by slipping a strip of card between the glass and the frame until the vibration stopped. The loose ends of the chains which supported the glass light shade above the cabinet eventually had to be cut off to stop them vibrating.

There were other vibrations, and the owner was asked to play the instrument whilst a search was made further afield. It is only by getting away from the source of generation that these sounds can be isolated. By presenting one ear to the instrument and one away from it, by walking in a gradually increasing semi-circle around the room, eventually the sound will be heard stronger in one ear than the other. Two such sounds were isolated in this case; the loose metal top of the venetian blind and a brooch in a glass ornament on the mantelpiece were the culprits.

It is unwise to ignore things which might be thought never to vibrate, such as the tail ends of strings leading from the bridge to hitch pins which have not one but three stringing braids fitted. It is because there is a strong tendency for them to cause extraneous sounds that this extreme fitting has been made.

The return of bracing timbers to the rear of pianos without the protecting cloth associated with older models places the soundboard at a distance from the external dimensions, allowing a space for all sorts of rubbish to lie against it and cause vibrations.

Noise in the region of the pressure bar can sometimes be eliminated by moving the offending string slightly. Beware of doing this if the hammers are very worn, as it will give a softer tone to the note in comparison with its fellows.

Lid studs used on grand pianos to centralise the top lid can make a sound out of all proportion to their size when they come loose. It is advisable to check grand hinge pins for vibration before raising the lid as these may not vibrate when the lid is raised.

Sounds can travel some considerable distance, and can be difficult to trace without the aid of someone else.

A room at the local Friends Meeting House was rented to a music teacher and he complained that the piano was making a terrible vibrating sound. A visit found that there was no way that any unusual sounds could be reproduced. A second visit was arranged at 9 a.m. on the morning of his normal visit to meet the teacher and get him to produce the sound he was complaining of. He explained the type of sound he had heard and proceeded to play the piano, producing the most dreadful noise.

It was not coming from the piano, though. It was coming from the wall heaters spaced round the room in lieu of central heating. These were switched on only when he occupied the room for the whole day. As the elements heated up, they expanded and were vibrating in sympathy with the piano.

He was advised that it was an electrician he needed, not a piano tuner.

As a young tuner, I was sent to see a Blüthner grand piano purchased by a local and very celebrated pianist. Three of the senior tuners had been and failed to trace a noise said to be emanating from the instrument, I had been sent only because I was a pianist and perhaps it was something pianistic. When he opened the door to this young fresh faced youth, the owner must have thought the firm were scraping the barrel this time, but he ushered me in civilly enough. At the far end of the lounge sat this large and imposing instrument. All the way to the house, I had been going over in my mind what could be wrong with it, feeling that if the senior tuners had not found anything wrong, then there must be nothing wrong and it must be something extraneous.

On playing it, there was what could only be described as a 'zinging' sound and I agreed with him that there was indeed something wrong. Having asked him to play it whilst I listened, I proceeded to walk around the piano with my left ear towards the instrument, progressing in ever widening semi-circles until I reached a position some 11 feet (335 cm) to the rear of him as he sat. It was at this point that I could hear it louder with my right ear which was directed away from him.

Edging slowly in the direction of the sound I came upon a large portrait of some long-dead ancestor of his and looking behind it could see a chain with a ring hanging from it. Reaching up, I held it still. 'It's gone!' he shouted, and proceeded to belabour the piano. As I released it, he said 'No, its back.' Returning to my tool box, I took my wire cutters and cut off the offending chain and ring, as he began once more to belabour the piano. When I showed him the cause, he could not believe it but at the same time, he could not reproduce the zinging sound again either. After celebrating in the company of his whisky bottle, I returned to the factory as my work had been suspended for the day whilst I attended to the problem.

By the time I arrived (we travelled by bus and tram in those days), he had been on the phone to the boss extolling my virtues as a troubleshooter. The following Friday, I received a bonus in my pay packet, equal to another week's wages.

Beaded top and bottom door panels if not securely fitted can cause dreadful buzzes but are easy to trace, as are split soundboards.

Not quite so easy was a school piano which was causing a problem of sympathetic vibration. When I arrived to test it, of course it wouldn't reproduce the sound but knowing that this particular music teacher would not complain if there was not a definite problem, I tested all the usual things that I have mentioned.

When looking at a problem like this it is best to start testing the instrument just as it is used, so I asked if there was anything close to or on the piano which was not here now. The teacher assured me that it was just the same, he never kept anything on the top of the piano which would vibrate.

Reaching down to test the castors to see if they were loose, I noticed that the wood block floor was very uneven. Moving the piano about 4" (10 cm) to the right, I let one castor hang over a dip in the floor and started to play it. The buzzing sound was quite pronounced. 'That's the sound,' he said, and I showed him what was causing it: the piano now stood on three castors, allowing the fourth to hang freely over the dip in the uneven floor and thus to vibrate. It transpired that the cleaner moved the piano each day to clean the room and replaced it in not quite the same position each time.

Curved reflectors on electric fires can catch the sound of a piano and rattle the metal guard fiercely. Glasses in cabinets which are just touching will vibrate if not given sufficient separation.

The main theme which comes from all of the above is, having satisfied oneself that the sounds are not coming from the instrument, involve the owner in playing the offending notes, and then start searching further afield, one ear towards the piano and the other away from it.

Sympathetic vibration is used in piano design to give a lift to the treble sound by means of Aliquot stringing. We have discussed the Bluthner patent and it might be prudent to look at other manufacturers' ideas. The Steinway Duplex patent and Kawai version are shown in Figures 116 and 117.

With the Blüthner patent, there is only one string vibrating at a time, except in the extreme treble above the damped strings. With both the Steinway and the Kawai patents, all of the strings are free to vibrate in sympathy.

The Bluthner patent allows precise tuning of the vibrating string for maximum effect. All others leave the effect to approximate tuning, this being effected by altering the position of the tuning bar which the string passes over between the soundboard bridge and the hitch pins. This is neither good or bad from the operational view but I have never felt that these approximations of tuning have the precise effect that the Blüthner patent has.

All strings will, if left free and undamped, vibrate in sympathy with other sounds of the same frequency or partials of them. We

TUNING: ADDING THE PROFESSIONAL TOUCH

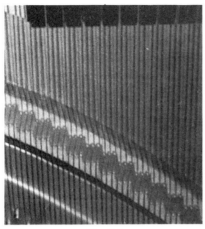

Fig. 116

Fig. 117

have seen that the Aliquot tuning of the Blüthner patent is effected by tuning it to a stretched octave above the companion note, until the Aliquot bridge transfers to the main bridge.

The tuning at this point is effected by tuning the Aliquot string the amount sharp that a stretched octave would be in the lower register. The slight amount sharp gives the note a lift to the sound thus produced.

Some years ago, this fact proved useful in solving a problem for a dear elderly lady who had already consulted all of the best tuners in the district. By the time I arrived on the scene, her grand piano was already perfectly in tune and I asked her to describe the problem. She started to play and described how the treble did not have the ringing sound she associated with the piano in times gone by. I played a similar piece to the one she had played, and I felt the same way about it myself. It was obvious that due to its age, the strings had fatigued, losing a lot of their elasticity and that their ability to vibrate in sympathy with the undamped treble strings was seriously impaired.

Plainly the instrument needed restringing. I told her why the sound had deteriorated but asked her to leave me to see if I could improve the situation.

From the point of the treble frame bar I proceeded to tune one of the strings of each note, just the amount sharp that a stretched octave would be, giving each note a slight edge. When I called her in to test the tuning, she was delighted with it. 'How have you done it?' she asked. 'I've put it out of tune,' I replied. 'Will anybody notice it?' she asked. 'Not if you don't tell them,' I replied. 'This must be our little secret then,' she said, and so it was for many years after. The Aliquot effect must not be overdone so that the piano is out of tune, but just enough to give the note an edge.

When I was first let loose on an unsuspecting public, I was sent to tune a piano which was so far out of tune that after going through the tuning once, I felt that I had to go through it again. Even after this second tuning, I retuned the treble again, until I felt satisfied with it.

On returning to the factory next day I was told that the customer had complained that the tuner had taken away the beautiful bell-like tone which the piano formerly produced. The indignation must have shown on my face as I explained the condition and the trouble I had taken over the instrument. The senior tuners were smiling. Dave Brunning took me to one side and told me to go back and make some excuse to retune the piano, this time taking one string of every note in the treble sharp until it jangled sufficiently out of tune to her satisfaction.

Making the excuse of having been advised by the factory that there had been something about her particular instrument that I had overlooked, I was allowed to retune it. When the piano was jangling to the amount sharp that I considered would satisfy her, I called her in to try it. 'That's better,' she said. 'You should have done it like that in the first place.' Muttering an apology, I crept away, another lesson learned.

Don't always expect praise for your efforts, no matter how hard you try.

Use of the small tuning head

Figure 118 is an excellent example of what happens when a medium tuning head is used on small wrestpins. The wire has been polished where the tuning head has touched the strings. In manufacture, the level of the coils has been varied in order that the strings do not touch neighbouring coils on their journey towards the pressure bar. When the medium head is used, it fits so far down the wrestpin that it touches the strings of notes already tuned.

The section shown was within the bearing area and my apprentice could not understand why his tuning was apparently not staying where he put it.

If we look at the first coil of wire on the wrestpin, some of them are quite shiny where the tuning head has polished the wire, as are various sections of wire where they extend towards the pressure bar and have been touched by the tuning head in the process of tuning a neighbouring note. All of these pressure points will destroy any tuning present if touched after being tuned and will nullify the efforts of the best tuner.

Fig. 118

The small tuning head only just clears the highest strings of the adjoining notes and a written reminder to this effect on the record card should be made for the benefit of the next tuner or even for oneself, as a constant note of the instrument's particular problem.

Whilst advocating the use of the small head, I must reiterate the warning of the converse effect of using the small head on large wrestpins and the danger of chewing off the corners by gripping the pin too high, which does not give enough purchase.

Electronic tuning aids

There have been many electronic tuning aids on the market over the past twenty-five years or so, and overall they have not made the impact upon the trade that they might have had if sufficient prototype testing had been done.

This does not say, however, that some models, namely R.B. Instruments ETA Mk. 6 and later the Vista II, were not extremely useful pieces of equipment. It would be wrong to say that they do not work, and if every instrument we had to tune were a 9' (275 cm) grand then tuning by this means would be simple, requiring only a modicum of manual dexterity and little experience to perform the task.

Grands of this length are comparatively rare and the majority are of much shorter length, although of possibly good quality, cannot produce correctly the range of partials that the larger grands cans.

What electronic tuning aids cannot take into account is inharmonicity, the inability of the strings of shorter length or lesser quality instruments to produce the particular partial correctly to which the aid is tuned. In a word, they cannot think; the thinking has to be done by the tuner.

The Vista II has just the right sensitivity of positive and negative indication to produce an excellent result.

If, as in all smaller instruments there has been a modification to string length, thickness or by excessive overwinding, then, when these strings are consonanced to the tuning aid, an incorrect reading will result. In cases like this, a search must be made to find a partial of another note which will consonance with the note being tuned.

Over a number of years, my partner and I have experimented and have found a set of partials which produces a reasonable result, but we always conduct ear tests to smooth the design discrepancies of each individual instrument.

One thing which this tuning aid has established is the tendency of pitch to creep in an upward direction in the area of the central frame break. In the central section of strings upward from the break for the first four or five notes, pitch can raise by as much as $1/4$ to $1/2$ hertz (cycle) at each tuning. In cases of damp this can be as much as 1 or even 2 hertz.

As these strings frequently come within the area of the bearing, care must be taken to ensure that the bearing does not creep in an upward direction resulting in an overpitched instrument. This phenomenon can be useful if the instrument being tuned is below pitch, by setting the bearing to the high point and allowing the bearing to creep upward until the pitch is correct. Take note of this when humidifying an instrument, keeping a constant check on the pitch.

Another extremely helpful use of this equipment is in distinguishing the echo sound in the treble as it will only consonance with the tuning frequency, not the extraneous vibration.

In the set of circumstances where the windings in the bichord section will not agree, the equipment will show the correctness of the tuning of each string and thus the incompatibility of the windings. Only time and practice brings the firm but gentle touch required to move the string only by the amount required and no more. The sensitivity of these tuning aids shows this heaviness up very quickly and is extremely useful in refining a tuner's touch.

By showing movement, pace and direction, the co-ordination of hand, ear and eye improves in a very short time ensuring early string stability by less movement.

In all usage of equipment of this kind, ear tests must be incorporated at every stage, the roll of the tuning aid being kept to that of useful aid.

The professional tuner expects to tune a string in a matter of seconds, making no more than five or six movements to a stubborn string. If there is any doubt in his mind, the string or note is left to rest as the tuning proceeds until the string or note becomes the 4th or 5th test interval, when another two or three attempts will be made. If this is still unsuccessful, it is left to rest again until it becomes the octave of the note being tuned, when a single final attempt is made. Further agitation of the stubborn string or note will only result in string instability.

Apprentices under tuning instruction always over-agitate the strings and are always surprised when the tutor tuner can pinpoint all the strings which have caused them problems. In spite of their efforts and because of this over-agitation, these strings are more out of tune than the rest.

When an instrument is badly out of tune the inexperienced tuner spends an unnecessarily long time setting the bearing, or laying the scale, only to find that by the time the tuning is completed, the bearing is wildly out again.

The experienced tuner lays a rough bearing starting on C and tuning all the Cs on the piano, then G and all the Gs and so on, proceeding around the bearing in this fashion raising the pitch to a point slightly above pitch, which allows for a slight compression due to the extra strain.

To show that this can be done in twenty minutes, I make a point of demonstrating it to my apprentices. I believe the record in the *Guinness Book of Records* is nine minutes, but the average tuner is content with twenty minutes. The effect is to place an even stress over the whole instrument, not too much in any one place.

By the time this is done, Mrs Smith will probably be bringing in the tea. If not, the pause to let the instrument rest will provide the hint that a short break is called for.

The second phase should concentrate on the bearing and then proceed down to the bass end. By this time, the treble will have sunk badly and a swift tuning in octaves to the top should smooth this out. Next, check the bearing, and proceed up to the top as normal.

There are variations to this theme, but all should be directed to a swift pitch raising spread over the whole instrument, followed by a rest and a tuning on pitch.

The pressures acting upon the instrument will then equalise with time. Any further agitation is a waste of time and money. The money is better spent on the succeeding tuning at least four times in the following year, which will bring with it the instrument stability which is so necessary to good tuning.

Playing-to-tuning ratio

In the 1920s and 1930s, when a new piano was purchased from my firm four free tunings were given, to be made at three-monthly intervals during the first year. This enabled the tuner to establish instrument stability. According to use, the customer would probably reduce this in succeeding years to three or two, which, if the playing-to-tuning ratio was correct would be quite adequate.

To show something of the playing-to-tuning ratio, we have considered the concert tuning frequency. Probably the next in frequency are theatre pianos.

My local theatre, the Floral Pavilion in New Brighton, can have a show on for five week days and a one-night stand on Sunday. The tuning for the weekday show would be completed on Monday before the band call at 10 a.m. The tuning for the Sunday show would be completed on Sunday morning at the same time. These tunings were shared between my partner and myself and it might be worth noting before committing oneself to such work, that classical concerts take place on Saturdays and Sundays, requiring attention from the afternoon until the evening interval check.

Next in line come club pianos used two or three nights per week, usually tuned monthly. If they are used perhaps one night per week they are usually tuned every three months, as are most music teachers' and most professional musicians'. All the above frequencies must take into account the pianists' own desire for quality of tuning.

The amount of use school pianos get justifies tuning four times yearly, and up to around 1950 this was the case. The financial reductions which have taken place since do not allow this standard. When local management comes into operation, there maybe schools of musical quality who will return to perhaps three tunings yearly, but this remains to be seen. Overplaying and undertuning results in instrument instability.

A case in point happened some years ago when my partner advised me of a problem he had with a Bentley upright which would not stay in tune with the usual twice-yearly tuning. Each time, the instrument was down in pitch and the tuning was in a chaotic state. Although the customer was not complaining unduly, my partner's efforts to humidify, checking that the central heating was kept low, did not seem to have any beneficial effects. He could find no structural reason for the problem.

When the piano was due for its next tuning, I arranged to call and check it over, and found that indeed it was in a bad state. There was no structural reason for this, so I asked the owner the amount of use it was getting. She replied, 'I play and the children play.' The children were only up to grades two and three.

A form was left with her with a space for each day, to put down exactly how much use the piano received over a period of three months. I found that she played for seven hours per week, and what I had fondly thought of as her own children consisted of forty pupils each receiving a half-hour lesson, making a total of twenty-seven hours per week.

When asked why she had not mentioned that there were so many pupils, she said, 'I didn't think that their little fingers could do much damage.' When I explained to her that their playing plus her own, coupled with the demonstrations she was giving in the treble, amounted to some considerable usage, she agreed to increase the tunings to four times per year. After a period of two years the instrument stabilized and caused no further problem.

At this point, it might be wise to mention the particular pattern of wear associated with music teachers' pianos. By sitting to the right (treble) of their pupils and demonstrating in the area of the treble break, where the hammer felt is getting thinner, their instruments receive considerably more wear at this point from the combined efforts of their pupils and themselves. The tuning in this area also suffers badly. All music teachers should be encouraged to sit to the left (bass) of their instruments, demonstrating in a more central position where the felts are thicker.

Piano houses which place a label on their instruments advising that they should be tuned twice yearly, give some owners the impression that this is all that will ever be required regardless of the amount of use they are subjected to, and resist all efforts to increase the frequency needed from excessive use.

A new modern instrument of reasonable quality played for five or six hours a week should hold in tune for six months. On older instruments, frequency can vary according to age, wear, humidity, wrestplank condition, central heating and use. Tuning should be adjusted to take these factors into account until stability is obtained.

The wrestplank of an instrument with random loose wrestpins can be sealed by humidity, but due to the different construction of older models pre-1940 and the glue used in construction, the drying effects of central heating can cause structural damage which no amount of humidifying can correct. An increase in frequency of tuning can relieve the constant stressing and restressing of older instrument which takes place at each tuning when it is allowed to go badly out of tune.

Many things which have been spoken about in this chapter eventually become second nature as experience increases, but by bearing them in mind at this early stage much time and heart-searching will be saved.

Keyboard temperaments

There are many keyboard temperaments which tuners can be called upon to set, including the usual EQUAL TEMPERAMENT, any of the dispositions of PYTHAGOREAN INTONATION and also the many MEANTONE temperaments. The tuning being chosen to complement the music of a particular historical period.

Equal temperament

Many tuners profess only a very basic knowledge of the tests for accuracy which can be applied to equal temperament tuning but it is only these tests which will prove the accuracy of one's work. Just as the proof of the pudding is in the eating, so the proof of the tuning is in the testing. If one's tuning does not stand up to the testing, then either change the chosen method, or improve the quality of the tuning.

All tuners are instructed in what professes to be equal temperament. The methods of producing the end product vary considerably. Some are standard methods, others are not, and as I am often asked about by own chosen method, we might as well make a start there.

This method is a standard one which has been taught over a great many years with success. I do not consider myself to be the world's best tuner, but by giving constant attention to

    accuracy
    consistency
    string stability
    instrument stability
    humidity
    playing-to-tuning ratio

the overall results of my work are better than average.

Before launching into a discussion of tuning, it would be prudent to define the terminology and meaning of an interval.

The interval

An octave musically speaking has no middle, that is to say, it is divided into unequal (but balanced) divisions of a fourth and fifth, or a third and sixth. This is tied up with the mechanics and mathematics of the diatonic scale and as this book is restricted to the practical application of tuning, any calculations have been worked out to keep the theoretical side to B-Tec level.

- An *interval* then, is the *distance* between any two notes.
- A *pure (perfect)* interval is an interval which has *no beats*.
- A *tempered interval* is an interval which is:

    lesser or greater than pure
    smaller or larger than pure
    narrower or wider than pure.

All of these terms have been used to describe the adjustment of the *distance* between any two notes to make a tempered interval.

Although an octave has no beats, it is greater than pure when it is stretched. A Major 3rd/Minor 3rd, 6th interval or any inversion of these had a designated speed of beating according to the type of tuning, i.e. TEMPERED/PYTHAGOREAN or MEANTONE, some using pure or almost pure 3rd and 6ths, others a mixture of tempered and untempered intervals.

In setting out a short tuning cycle, the designation of < to mean *down to*, and > to mean *up to* has been used in order to avoid confusion as to the direction of progression and designation of interval, e.g. C < F 5th interval as opposed to C > F 4th interval.

My reason for choosing the following method is that it provides the earliest countable test intervals, 7 or 8 beats per second which is my most practised judgement of speed. These speeds are the highest which can be accurately counted; speeds higher than this must be judged from this point by the progression of speed, faster (upwards) and slower (downwards).

In the tables on pp 146/147 each tuned interval being followed where applicable by its associated test interval(s). Most tuners have a British standard C (A 440) tuning fork and the tuning cycle commences with the consonacing of middle C to this fork, progressing down to lower F. Follow along each horizontal line, tuning and applying the test intervals as they occur, until the cycle has been completed.

Short tuning cycle of 4th and –

(> = *up to*) (< = *down to*) (# indicates the sharps)

SCALE NOTES	INTERVAL	BEAT RATE PER SEC.	BEAT RATE PER 5 SEC.
C < F	5th	0.590 (0.6)	2.95 (3.0)
C < G	4th	0.886 (0.9)	4.43 (4.5)
G > D	5th	0.664 (0.65)	3.32 (3.5)
D < A	4th	0.995 (1.0)	4.98 (5.0)

Pause here, in order to establish the foundation of the scale.*

A > E	5th	0.744 (0.75)	3.72 (3.75)
E < B	4th	1.116 (1.0)	5.58 (5.5)
B < F#	4th	0.838 (0.8)	3.94 (4.0)
F# > C#	5th	0.625 (0.6)	3.125 (3.0)
C# < G#	4th	0.941 (0.95)	4.71 (4.75)
G# > D#	5th	0.702 (0.7)	3.51 (3.5)
D# < A#	4th	1.053 (1.0)	5.27 (5.25)
A# < F	4th	0.790 (0.8)	3.95 (4.0)

There is no reason why a tuning cycle should not commence with the tuning of A 440, the cycle being completed in the same way as when starting on C, but by starting all tunings in this manner, it will be seen that when setting all other temperaments they can be completed in the same way without having to learn a different method of tuning for each one.

It is no use learning a tuning method which tunes only the 3rd and 6ths, when some tunings have 3rd or 6th intervals which are either pure, or nearly so, demanding pure 4ths and 5ths, which are not receiving the attention they must have.

– 5th intervals with tests

TYPE OF INTERVAL	MAJOR 6TH TEST INTERVAL	MAJOR 3RD TEST INTERVAL
Less than Pure	–	–
Greater than Pure	–	–
Less than Pure	F > D 7.925 (8.0)	–
Greater than Pure	–	F > A 6.930 (7.0)
Less than Pure	G > E 8.895 (9.0)	C > E 10.382 (10.25)
Greater than Pure	–	G > B 7.778 (7.75)
Greater than Pure	–	–
Less than Pure	–	A > C# 8.732 (8.75)
Greater than Pure	–	G# > C 8.24 (8.2)
Less than Pure	F# > D# 8.395 (8.5)	B > D# 9.798 (9.75)
Greater than Pure	–	F# > A# 7.343 (7.5)
		A# > D 9.25 (9.25)
Less than Pure	–	–

Note: The last interval A# < F is the untuned test interval and is the critical test of the correctness of the tuning progression throughout the gamut of the scale.

*Having reached this stage, I do not let my apprentices proceed further until these intervals are established. We have the three earliest and most countable test intervals within the gamut of the scale. If these are established at this early stage, the remainder of the bearing will be set on an accurate foundation.

The tuned interval A<D (1.0)
  The 6th interval F>D (8.0)
  The 3rd interval F>A (7.0)

It can be seen that all of the above intervals depend on the tuner's judgement and the result is what is termed rough tuning. It is not very often that a scale is completed without discernible error and tests have to be used to *fine tune* the scale to our satisfaction.

Testing the progression of major 3rds/6ths within the bearing

F > A Maj. 3rd	(7.0)	F > D Maj. 6th	(8.0)
F# > A# „ „	(7.5)	F# > D# „ „	(8.4)
G > B „ „	(7.75)	G > E „ „	(8.9)
G# > C „ „	(8.25)		
A > C# „ „	(8.75)		
A# > D „ „	(9.25)		
B > D# „ „	(9.75)		
C > E „ „	(10.25)		

The value of testing the progressions lies in their ability to show if there is any discrepancy in the speeds, it can be seen that the major 3rd should progress smoothly upwards from 7.0 to 10.25, every interval being progressively faster than its neighbour, no sudden jumps in speed or slowing down. The major 6th intervals progressing from 8.0 to 8.9 in the same way.

Testing the progression of minor 3rds within the bearing

F   > G# (9.50)
F# > A   (10.00)
G   > A# (10.50)
G# > B   (11.25)
A   > C   (11.75)
A# > C# (12.50)
B   > D   (13.25)
C   > D# (14.00)
C# > E   (15.00)

Note the increase in speed from 9.5 to 15.0. It is not possible to count these speeds, only to develop a practised judgement of the progressional speed.

If the above tests are made and errors are perceived, then there must be a procedure established to find and correct the incorrect intervals. The above tests only show that errors exist; what they don't show is which notes are causing them.

Searching the bearing for error
Finding the note to adjust when a progression of 3rds or 6ths does not show the graded increase in speed required of them must be a process of elimination. That is why it is often referred to as smoothing.

First, look at the intervals which most nearly beat at eight beats per second, time spent counting and perfecting one's judgement of this speed is never wasted.

Those who have a quartz pulsar digital watch will notice that an exact second is displayed by a pulsing beat indicating the space of one second. Count ONE – TWO – THREE – FOUR – FIVE – SIX – SEVEN – EIGHT, saying the complete words in the space of one second until you can say it exactly within the specified time. This is what I make my apprentices do every morning until I am satisfied that they can do it correctly.

Major 3rds	*Major 6ths*
G > B (7.75)	F > D (8.0)
G# > C (8.25)	F# > D# (8.4)

Taking a consensus of the intervals which are considered to beat correctly and judging from these the degree of error of the others makes a good starting point.

Let us take the case of the F needing adjustment. The 6th interval F > D will beat too slow or too fast according to whether the F is too sharp or too flat, as will the third interval F > A.

The minor third F > G# will also be out in the same direction. If the beat rates are too slow, the F needs to be adjusted in a downward direction to increase the beat rate of all three intervals, showing that the 5th interval C < F needs to be made greater.

Retune the F and retest all three intervals. It may be that only two of these intervals will show the need for this adjustment but this only shows that another note is also slightly out. Once again, take a consensus of the intervals. If the opposite obtains, then take the opposite action.

REMEMBER that the smallest adjustment of the 5th interval will bring about a much greater change in the beat rates of the 3rds and 6ths.

Remember also at this stage that there was no test for the F#

until the tuning cycle was completed and that any error might be compounded due to this fact.

Test F# in the same way, test the speed of the 6th F# > D#, the major 3rd F# > A#, and the minor 3rd F# > A and from the direction indicated judge the direction that the interval F# > C# needs to be adjusted.

In the centre of the bearing, notes such as A# can be tested against the major third above, A# > D, the major 3rd below A# < F#, the minor 3rd above A# > C# and the minor 3rd below A# < G. Proceed through the bearing in this fashion, smoothing out errors until satisfied that the scale has been set to one's satisfaction.

From the bearing which has just been set, it is now possible to extend it upwards or downwards. For the sake of simplicity, I will take the upward direction to the treble end commencing with the 5th interval A# > F.

We now have our first octave F > F to tune testing with

C# > F Major 3rd 10.997 (11.0)
G# > F Major 6th   9.424 (9.5)
D > F Minor 3rd 15.850 (16.0)

Make sure that the speeds are progressive from the intervals already tuned and that the octave has no beats.

The laws of octave balance

There are many ways of proving the truth of an octave and the laws governing them are best understood at this stage.

In general, we do not call the 5th interval or the 4th interval major and minor respectively, but, within an octave, the 4th is in fact a minor interval to its upper 5th.

Commencing with low F of the bearing we have just tuned, F > A# (4th) is the minor interval to its upper 5th A# > F and if we look at the beat rates, we find that the beat rates are exactly the same (0.790), therefore, if correctly tuned, they should balance.

If we then take the minor 3rd F > G# (9.5) and balance it against its reciprocal upper major 6th G# > F to make an octave, we find that once again, the beat rate should be (9.5), therefore, they should also balance if the octave is tuned correctly.

From this, we can then safely say

Within any octave, lower minor intervals must balance their reciprocal upper major intervals, beating at the same rate.

TUNING: ADDING THE PROFESSIONAL TOUCH

Commencing with the low F of the bearing we have just set, the following tables illustrate the relationships between these intervals, progressing in an upward direction, showing also, that when the series starts again in the next higher level, that the beat rates in all cases have doubled.

MINOR 3RD INTERVALS	RECIPROCAL UPPER MAJOR 6TH INTERVALS
F > G# (9.5)	G# > F (9.5)
F# > A (10.0)	A > F# (10.0)
G > A# (10.5)	A# > G (10.5)
G# > B (11.25)	B > G# (11.25)
A > C (11.75)	C > A (11.75)
A# > C# (12.5)	C# > A# (12.5)
B > D (13.25)	D > B (13.25)
C > D# (14.0)	D# > C (14.0)
C# > E (15.0)	E > C# (15.0)
D > F (16.0)	F > D (16.0)
D# > F# (16.8)	F# > D# (16.8)
E > G (17.8)	G > E (17.8)
F > G# (19.0)	G# > F (19.0)

MINOR 4TH INTERVALS	RECIPROCAL UPPER MAJOR 5TH INTERVALS
F > A# (0.790)	A# > F (0.790)
F# > B (0.838)	B > F# (0.838)
G > C (0.886)	C > G (0.886)
G# > C# (0.941)	C# > G# (0.941)
A > D (0.995)	D > A (0.995)
A# > D# (1.053)	D# > A# (1.053)
B > E (1.116)	E > B (1.116)
C > F (1.180)	F > C (1.180)
C# > F# (1.250)	F# > C# (1.250)
D > G (1.328)	G > D (1.328)
D# > G# (1.404)	G# > D# (1.404)
E > A (1.488)	A > E (1.488)
F > A# (1.580)	A# > F (1.580)

This situation can be reversed. If we place the major interval in the lower position and the minor interval in the upper position, we find that the major interval beats at exactly half the speed of its reciprocal upper minor interval.

MAJOR 5TH INTERVALS	RECIPROCAL UPPER MINOR 4TH INTERVALS
F > C (0.590)	C > F (1.180)
F# > C# (0.625)	C# > F# (1.250)
G > D (0.664)	D > G (1.328)
G# > D# (0.702)	D# > G# (1.404)
A > E (0.744)	E > A (1.488)
A# > F (0.790)	F > A# (1.580)

MAJOR 6TH INTERVALS	RECIPROCAL UPPER MINOR 3RD INTERVALS
F > D (7.925)	D > F (15.580)
F# > D# (8.395)	D# > F# (16.790)
G > E (8.895)	E > G (17.790)

MAJOR 3RD INTERVALS	RECIPROCAL UPPER MINOR 6TH INTERVALS
F > A (6.93)	A > F (13.86)
F# > A# (7.34)	A# > F# (14.68)
G > B (7.78)	B > G (15.56)

From this, we can safely say,

within any octave, lower major intervals will beat at half the rate of its reciprocal upper minor interval.

With so many tests of the exactitude of an octave, there should be no excuse for mistuning it.

It can be seen at this stage, that beat rates are reaching the stage that counting is just not possible and it is more a practised judgement of the speed progression; no sudden increase or decrease in speed but a smooth progression as has already been produced within the bearing.

When the A above middle C has been reached (A440), we have yet another test which can be applied, THE 10TH INTERVAL.

The 3rds at this stage are beating very fast but can be slowed down to manageable proportions by *inverting* the 3rds and making them into 10ths.

i.e. Low F > A (440) major 10th (an octave + a major 3rd)
Low F# > A (440) minor 10th (an octave + a minor 3rd)

Speed test
The speed of the 10th interval is reflected in the lower 3rd interval e.g.

> F > A (6.93) major 3rd interval should beat at the same rate as F > A (440) major 10th interval, thus halving the rate of the fast beating upper 3rd interval top F > A (440) (13.86)

This test can be continued, reducing the beat rate by means of the 10th interval until the need for the use of the double octave + a major 3rd is reached. By this time, the need for octave stretch is approaching.

Octave stretch
If the tuned intervals are correct and the speed increases of the test intervals progressive, then the higher octaves should be correct. However, in the upper reaches of the treble, all of these intervals become increasingly difficult to judge and the *sound* of the octave becomes increasingly important.

Even if the octave beats are tuned correctly at this stage, the overall sound would be flat and some adjustment must be made to make this acceptable to the listener. This does not mean that the octaves must be sharpened to the extent that they are beating sharp. Many tuners err on the side of sharpness in the belief that musicians prefer this, and indeed many musicians do not notice as much sharpness as they would instantly notice flatness.

The lift that a correctly stretched octave gives to the overall sound of the instrument is one of the pleasures I get when listening to a performance whilst remaining to retune again during the interval.

Everything I said about the lift that the Blüthner Aliquot stringing gives to a note applies to the stretched octave.

The best explanation of octave stretch was given to me by my guide and mentor Dave Brunning and I feel that I can give no better explanation. An octave is said to be stretched, when the string reaches the point when it starts to 'zing' providing the note with an edge, just before beating sharp. We all experience the increasing 'zing' just before the string bursts into oscillation. Practice listening to this until the point becomes fixed in the mind and can be sensed as much as heard.

Hearing focus
A great deal of rubbish is spoken about the ability of blind piano tuners and musicians to hear things that other people cannot. THIS IS NOT SO.

Although hearing varies, what happens is that many people whose jobs, professions or even disabilities entail the use of sound, develop the ability to focus their hearing to better effect and interpreting what they hear by their knowledge, practice and understanding of what the sounds mean. This is what tuners must use when tuning the treble and bass ends of instruments in order to grade the expanded scale throughout the compass of the instrument.

Bass tuning
Extending the scale towards the bass end brings us very quickly to the problem of inharmonicity, the inability of a particular string due to modification to produce a correct partial series.

It is now that the more tests that can be applied to the tuning of a particular note, the easier it will be to make the necessary compromises.

With all tuning below the bearing, the TUNE AND SEARCH method is probably the best. The method varies from instrument to instrument according to quality and physical size, as does the response to the various tests. In including all tests which can be applied, it must be remembered that a miniature piano will not give the response to them that a large grand will.

It is not possible to give beat rates for the lower bass intervals as these will also vary, and emphasis must be placed on the natural progression of each individual instrument, compromises being made accordingly.

To proceed downwards from the bearing, we commence as follows:

Tune B < E (5th interval)
Testing with E < E (octave balance tests)
A   < E (4th interval)
G# < E (3rd interval)
C# < E (6th interval)
G4 < E (10th interval)

It is possible to make all the necessary compromises by listening to the reducing beat rate of all the test intervals down to possibly A2 on a large grand but quite possibly only C2 on a miniature piano. When a stronger beat rate is required lower down, the method of using the double octave + a third can be employed with a good measure of success, especially on a large grand.

NOTE TO BE TUNED	TEST NOTE
C 2	E 4
B 2	D#4
A#2	D 4
A 2	C#4
G 1	C 4

and so on down to A1. By referring back to the bearing all the time (4th octave), the extended scale will always remain in balance.

By this method, the reducing beat rate progression is monitored from a standard point.

In order to get used to these progressions, the speeds can be tested from above this range and will show the rate of slowing down as they progress lower down the scale. Pay great attention to the reducing speed of the 10th intervals as this is a frequently played interval musically and if left beating too strongly is a frequent cause of complaint.

It has been frequently said that it is impossible to apply all of the above tests on every instrument at every tuning, and this is quite correct. Indeed, it would be a waste of time on any instrument until all of the following conditions have been achieved.

- Accuracy
- Consistency
- String stability
- Instrument stability
- Humidity
- Playing-to-tuning ratio.

The fine tuning and the refining of the scale can then be achieved with the resultant increase in quality associated with *craftsmanship* as opposed to *workmanship*.

Pythagorean tuning

In my consideration of equal temperament, I have used only the # to indicate a sharp and designate the semitones between the naturals (the black notes between the white notes) of the scale, as this is the indication used by many piano manufacturers and is often punched into the wrestplanks of their instruments.

Musically however, the sound of a sharp should be different from the sound of a flat, and because of this some early keyboard instruments have the black notes divided into two, giving two separate sounds, one for the sharp and the other one for the flat. This arrangement gives the musician a closer rendering of the musically correct sound, which in turn gives a more pleasing rendering of a particular range of music. The method was both difficult and cumbersome and has not survived.

Equal temperament makes the sound of the sharp and the flat the same, thus allowing the modulation through all keys at the expense of the less musically accurate sound.

Pythagorean or Meantone tunings, whilst allowing a more pleasing rendering to the ear, have the disadvantage of limiting the musician and composer to the modulation in and out of closely related keys, restricting modulation into other keys which sound out of tune.

The increasing interest in early music has prompted many calls for tuners to tune harpsichords, clavichords and spinets in the various dispositions of Pythagorean intonation and Meantone temperaments.

Most instructions I have read on the subject have been given in musical terms for the benefit of musicians, and as many tuners are not musicians requests for information on the subject in language they can understand has prompted me to put this down in piano-tuning terms together with some idea of the associated musical period and the uses to which they can be put.

Pitch

For a concert of early music, my apprentice Jeremy and I were asked to tune to a pitch of A415. Jeremy's reaction was, 'How do we do that?' This is not an unusual request for such concerts, which are often conducted with stringed instruments or vocal works alone.

Pitch has undergone wild variations over the years, and early music still sounds best at this pitch. However, most other instruments of a fixed pitch are still designed to be played at A440, and if these instruments are to be used the keyboard tuning must also be pitched at A440. Pythagorean tuning can be effected at any pitch to suit other wind instruments or organs.

If we take A440 and divide it by 1.0594631 (12th root of 2), we get 415.3046953, the frequency of G sharp/A flat below A440, or if C, which is probably the tuning fork most tuners carry, is the calculation note, the frequency of B, one note below.

It is unlikely that a tuning fork of this pitch will be carried, but most music societies requiring this pitch regularly will have one available.

Before engaging upon a concert such as this, it is essential to check the pitch required and the availability of the appropriate tuning fork.

If caught out on this point, it demands a certain amount of mental agility to tune the A sharp/B flat to the A440 fork or the C sharp above middle C to the C fork, proceeding around the gamut of the scale one note high; not impossible but a little difficult.

From now on, a distinction will have to be made between a sharp and a flat. (#) will indicate the *sound* of a sharp and (♭) will indicate the *sound* of a flat.

Earliest Pythagorean intonation

For the earliest extant keyboard music, in the Robertsbridge Fragment, or an exploration of plainchant, organum and early Gothic polyphony, the following Pythagorean intonation, with its pure 4ths and 5ths will give the best results.

This disposition having two flats and three sharps will favour music which is written with key signatures having this balance of sharps, flats and accidentals.

Note that there is no A♭, but we do have a G# instead, therefore music written in the key of A♭ or music which uses this note will sound out of tune. Tune the following cycle of 4ths and 5ths *pure* (no beats).

    C < G 4th interval
    C < F 5th interval
    F > B♭ 4th interval
    B♭ > E♭ 4th interval

As we wish to leave the Pythagorean 'wolf' 5th between E♭ and G#, we must return to G and go around the other side of the gamut of the scale, all intervals tuned *pure* (no beats).

    G > D 5th interval
    D < A 4th interval
    A > E 5th interval
    E < B 4th interval
    B < F# 4th interval
    F# > C# 5th interval
    C# < G# 4th interval

The untuned interval between G# and E♭ is left less than pure beating at 5 beats per second. This is the howling Pythagorean wolf.

Before beginning tests of the scale, it is best to extend the scale in both directions to provide a larger test area which will encompass the reciprocal 4ths and 5ths to the 5ths and 4ths already tuned.

Extend the scale downwards by tuning pure 5ths, testing to see that their reciprocal 4ths and octaves are also perfect as follows:

    Tune B < E 5th interval pure
    Test B > E 4th interval pure (already tuned)
    Test E < E octave interval pure
    Tune B♭ < E♭ 5th interval pure
    Test B♭ > E♭ 4th interval pure (already tuned)
    Test E♭ < E♭ octave

Continue in this manner until the 5th interval E♭ < G# is reached. This once again is the wolf 5th and by the law of balanced octaves, must beat at the same rate as its reciprocal upper wolf 4th interval, although *slower* than in the bearing, whilst the octave is perfect.

If we check back to our bearing, we will find that the wolf 5th interval G# > E♭ now has a reciprocal wolf 4th interval G# < E♭ which by the laws of balanced octaves should beat at the same rate as its upper 5th. As we continued downwards by tuning pure 5ths only, the wolf 4th will not have been noticed.

From the bearing, proceed upwards in the same manner of tuning pure 5ths and octaves, until the 5th interval G# > E♭ is reached, once again, the 5th interval should beat at the same rate as its reciprocal lower 4th, although *faster* than in the bearing, whilst the octave remains beatless.

There is now a range of the instrument on which tests can be conducted. For a greater understanding of the tuning, we must now look at the effect that this has had on the major 3rds and 6ths.

The following major 6th intervals should be almost pure.

    G# > F
    C# > B♭
    F# > E♭

Starting with the lowest 6th interval tuned, G# > F, which should be *almost* pure, the intervals between will have an increasing beat rate from about 5 to 15 as they progress up the keyboard as far as has been tuned.

The following major 3rd intervals within the bearing should be *almost* pure.

F# > B♭
G# > C
B > E♭
C# > F

The intervals between, whilst beating faster than in tempered tuning, have an increasing beat rate as they progress up the keyboard as do the 6th intervals.

NO ATTEMPT should be made to alter the beat rates of these intervals. It is the purity of the 4th/5th intervals and the octaves which is the important factor.

In extending the tuning to the ends of the keyboard, tune for the purity of the 5ths and octaves, testing with the 10th intervals, remembering that those 3rd intervals which have a beat will produce a 10th interval with the same beat.

The major 3rds which are almost pure, will produce a 10th interval which will be almost pure.

Chasing the wolf

The wolf can be chased around the scale or bearing to produce other Pythagorean dispositions.

For certain pieces in the 15th-century *Buxheim Organ Book*, the three sharps have to be tuned differently.

If the preceding disposition has been set, it is possible to alter it as follows:

Start on E♭ < G#, adjusting the G# until pure with E♭, adjust the interval G# > C#, adjusting the C# until pure with G#, adjust the interval C# < F#, adjusting the F# until pure with C#.

This readjustment of the three sharps now places the wolf between F# and B (4th interval). The same alteration must be completed for the other octaves already tuned.

This disposition of Pythagorean intonation has been referred to by many early Renaissance theorists such as Henry Arnaut and John Hothby.

The system renders almost pure those major thirds which contain one sharp e.g.

D > F#
A > C#
E > G#
B > D# (note that we now have D# instead of E♭).

The following major 6th intervals are almost pure also.

A > F#
E > C#
B > G#

This imbalance of sharps over flats will favour music with key signatures containing sharps to the detriment of key signatures containing flats.

To set the above disposition from scratch, tune a series of pure 4ths and 5ths as follows:

C < G 4th interval pure
C < F 5th   "   "
G > D 5th   "   "
D < A 4th   "   "
A > E 5th   "   "
E < B 4th   "   "

As we wish to leave the wolf between F# and B, we must return to F and go around the other side of the gamut of the scale.

F > B♭ 5th   interval pure
B♭ > D# 4th   "   "
D# < G# 5th   "   "
G# > C# 4th   "   "
C# < F# 5th   "   "

The wolf now stands between F# and B, and is less than pure.

A third disposition, used for liturgical keyboard work of the early 15th-century Faenza Codex, calls for the wolf to be left between F# and C#. This only calls for the adjustment of F# to be made pure to B, leaving the wolf between F# and C#.

The major third which will be almost pure will now be as follows:

A > C# almost pure
E > G   "   "
B > D#   "   "
F# > B♭   "   "

The major 6ths will be as follows:

E > C# almost pure
B > G#   "   "
F# > D#   "   "

Remember that the intervening intervals will have a progressional beat rate from about 5 to 15 as they progress upwards as in our first scale. This does not change throughout all of the various dispositions.

Extending the scale throughout the keyboard must be completed in the same way as described for the first disposition.

The three foregoing dispositions of Pythagorean intonation will cover most of the music for which they are recommended. If however, there is any particular disposition which requires the wolf to be left in a different position, the preceding system remains the same.

Chase the wolf around the scale until it is in the required position, keeping to pure 4ths and 5ths. It can be seen that this scale can be altered by continuing the adjustments as follows:

make C# pure to F# leaving the wolf between C# and G#; make G# pure to C# leaving the wolf between G# and D# and so on around the scale until it gets back to its original starting position.

The disposition of the wolf must always be established from the musician/performer before tuning commences.

The obvious problems of the Pythagorean system showed the need of something better, and gradually over the years they produced alternative systems which solved some of the problems but not all of them.

Meantone tuning slowly replaced the Pythagorean system but for many years was used as an alternative to it. The musicians' problem of which system to use, or which system the composer intended, causes much argument and heart searching.

It must be understood that composers of these periods wrote pieces not just for a harpsichord but for a harpsichord tuned in a certain way, the details of which have in many cases not survived.

Meantone temperaments

Tunings with pure (perfect) major 3rds divided into two equal whole tones are called meantone tunings. The three REGULAR mean-

Scale (1)

SCALE NOTES	INTERVAL	BEAT RATE PER SEC.	TEST MAJOR 3RDS	TEST MAJOR 6THS
C < G	4th	1.35 Greater than pure		
C < F	5th	2.0 Less " "		
G > D	5th	1.65 Less " "		F > D 2.5 B.P.S. Greater than pure
D < A	4th	2.25 Greater " "	F > A pure	
A > E	5th	1.9 Less " "	C > E pure	G > E 3.0 B.P.S. Greater than pure
E < B	4th	2.5 Greater " "	G > B pure	
B < F#	4th	1.9 Greater " "		
F# > C#	5th	1.5 Less " "	A > C#	
C# < G#	4th	2.0 Greater " "		

As we wish to leave the wolf between G# and E♭, we must now proceed around the other side of the gamut of the scale.

F > B♭	4th	1.5 Greater than pure		
B♭ > E♭	4th	2.35 Greater " "	B♭ > D pure	

Notice at this stage, that we have only five pure major test 3rds and two major test 6ths. This is one of the main reasons for extending the scale once the bearing has been set. If the setting of the 4ths and 5ths can be looked upon as the rough or basic setting, and the purity of the 3rds are satisfactory. The 6ths having been checked for the correct beat rates. The minor 3rds can be tone tunings to be considered will cover most requirements, but there are also IRREGULAR meantone tunings to be considered later.

To obtain pure major 3rds, eleven 4ths and 5ths are tuned, in order to leave an extremely large wolf 5th, very much greater than pure. It is therefore essential that its position is established before tuning commences.

Scale (1) leaves it between G# > E♭.
Scale (2)   "   "   " A♭ > C#.
Scale (3)   "   "   " B♭ > D#.
Scale (1) produces 3 sharps F#, C#, G# and 2 flats B♭, E♭ (no A♭)
Scale (2)   "   2 sharps F#, C# and 3 flats B♭, E♭, A♭ (no G#)
Scale (3)   "   3 sharps F#, C#, D# and 2 flats B♭, A♭ (no E♭).

It can be seen from the above that the differing settings will affect the key signatures of the music which will sound better or worse according to the setting chosen, settings 1 and 3 favouring the sharp key signatures and setting 2 favouring the flat key signatures.

It must always be left up to the musician/performer to make the decision as to which setting will complement the music they have chosen to play.

The minor thirds can be tested for the following beat rates:

F# > A 3.0 beats per second
G > B♭ 3.5   "   "   "
G# > B 3.75 "   "   "
B > D 5.0   "   "   "
C > E♭ 5.5   "   "   "
C# > E 5.75 "   "   "

It will be noted that there are gaps in the progression of all test intervals due to the presence of the wolf 4th/5ths.

If we think about what we have done to our normal tempered scale, we will see that we have made the 4ths and 5ths more rough than usual, some of the major 3rds, pure, some remaining very rough. The minor 3rds are less rough than usual, beating much slower than in tempered tuning.

Remembering the law of balanced octaves, it must be realised that every wolf 5th must have a reciprocal wolf 4th in order that its octaves balance. As we do not get a smooth and continuous progression of 4ths and 5ths due to the wolf, we do not get a smooth and continuous progression of 3rds and 6ths either.

Extend the scale, tuning by octaves and 5ths, testing where applicable with the pure 3rds, balancing the octaves with their reciprocal 4ths and 5ths until the wolf is reached in either direction.

There is now a much greater range set which will allow a closer testing of the scale. All tests start at low A, the second A below middle C. If we play the minor 3rd A > C and balance it against its reciprocal upper major 6th C > A, by the law of balanced octaves they should beat at the same rate as in the following table.

MINOR 3RDS	RECIPROCAL UPPER MAJOR 6THS	BEAT RATE PER SEC.
A > C	C > A	2 to 3
B > D	D > B	v
C > E♭	E♭ > C	v
C# > E	E > C#	v
D > F	F > D	v
E > G	G > E	v
F# > A	A > F#	v
G > B♭	B♭ > G	v
G# > B	B > G#	v
A > C	C > A	4 to 6

If we test the progression of major 3rds which are now pure, against their reciprocal minor 6ths, we find that they should also be pure (if you double the beat rate of zero, you will still come out with a beat rate of zero).

MAJOR 3RDS	RECIPROCAL UPPER MINOR 6THS	BEAT RATE PER SEC.
A > C#	C# > A	Pure
B♭ > D	D > B♭	ʺ
C > E	E > C	ʺ
D > F#	F# > D	ʺ
E♭ > G	G > E♭	ʺ
E > G#	G# > E	ʺ
F > A	A > F	ʺ
G > B	B > G	ʺ
A > C#	C# > A	ʺ

If the above major 3rds are converted into major 10ths, they should still of course be pure.

We now look at the progression and balance of the (minor) 4ths against their reciprocal upper (major) 5ths.

TUNING: ADDING THE PROFESSIONAL TOUCH 165

MINOR 4THS	RECIPROCAL UPPER MAJOR 5THS	BEAT RATE PER SEC.
A > D	D > A	1.125
B♭ > E♭	E♭ > B♭	v
B > E	E > B	v
C > F	F > C	v
C# > F#	F# > C#	v
D > G	G > D	v
E > A	A > E	v
F > B♭	B♭ > F	v
F# > B	B > F#	v
G > C	C > G	v
G# > F#	F# > G#	v
A > D	D > A	2.25

Chasing the wolf

To change setting (1) to setting (3), it is only necessary to adjust the interval G# > E♭, adjusting the E♭ to make an interval with G#, two beats less than pure, which will in turn make E♭ into D#, leaving the wolf between B♭ and D#.

This will of course make changes in our progressions and intervals. If we set this temperament from scratch, we will see the changes which come about with the adjustment of a single note such as this.

We can now proceed to test the minor 3rds for the following beat rates:

F# > A 3.0 Beats per sec.
G > B♭ 3.5 ,, ,, ,,
G# > B 3.75 ,, ,, ,,
B > D 5.0 ,, ,, ,,
C# > E 5.75 ,, ,, ,,

Note the similar jump in beat rates to the last scale.

Extend the scale as before and start tests from low A, the second A below middle C.

Scale (3)

SCALE NOTES	INTERVAL	BEAT RATE PER SEC.		TEST MAJOR 3RDS	TEST MAJOR 6THS
C < G	4th	1.35	Greater than pure		
C < F	5th	2.0	Less		
G > D	5th	1.65	Less		F > D 2.5 Greater than pure
D < A	4th	2.25	Greater " "	F > A pure	
A > E	5th	1.9	Less " "	C > E pure	G > E 3.0 Greater than pure
E < B	4th	2.5	Greater " "	G > B pure	
B < F#	5th	1.9	Greater " "		
F# > C#	5th	1.5	Less " "	A > C# pure	
C# < G#	4th	2.0	Greater " "		
G# > D#	5th	2.0	Less " "	B > D# pure	F# > D# 2.75 Greater than pure

As we wish to leave the wolf between B♭ and D#, we must now proceed around the other side of the gamut of the scale.

F > B♭	4th	1.5	Greater than pure	B♭ > D pure	

Note that at this stage, we have 5 pure major test 3rds and 3 major test 6ths.

MINOR 3RDS	RECIPROCAL UPPER MAJOR 6THS	BEAT RATE PER SEC.
A > C	C > A	2 to 3
B > D	D > B	v
C# > E	E > C#	v
D > F	F > D	v
E > G	G > E	v
F# > A	A > F#	v
G > B♭	B♭ > G	v
G# > B	B > G#	v
A > C	C > A	4 to 6

If we now test the major 3rds and their balance against their reciprocal upper minor 6ths we should find that they too should all be pure. Remember that any major 3rd converted into a major 10th, should also be pure.

MAJOR 3RDS	RECIPROCAL UPPER MINOR 6THS	BEAT RATE PER SEC.
A > C#	C# > A	Pure
B♭ > D	D > B♭	″
B > D#	D# > B	″
C > E	E > C	″
D > F#	F# > D	″
E♭ > G	G > E♭	″
E > G#	G# > E	″
F > A	A > F	″
G > B	B > G	″
A > C#	C# > A	″

We can now look at the progression and balance of the (minor) 4ths against their reciprocal upper (major) 5ths.

| | RECIPROCAL | |
MINOR 4THS	UPPER MAJOR 5THS	BEAT RATE PER SEC.
A > D	D > A	1.125
B > E	E > B	v
C > F	F > C	v
C# > F#	F# > C#	v
D > G	G > D	v
D# > G#	G# > D#	v
E > A	A > E	v
F > B♭	B♭ > F	v
F# > B	B > F#	v
G > C	C > G	v
G# > F#	F# > G#	v
A > D	D > A	2.25

The minor 3rds should now be tested for the following beat rates.

F# > A 3.0 beats per sec.
G > B♭ 3.5   ,,   ,, ,,
A > C 4.0   ,,   ,, ,,
B > D 5.0   ,,   ,, ,,
C# > E 5.75   ,,   ,, ,,

The scale should now be extended upwards and downwards as before.

| | RECIPROCAL | |
MINOR 3RDS	UPPER MAJOR 6THS	BEAT RATE PER SEC.
A > C	C > A	2 to 3
B > D	D > B	v
C# > E	E > C#	v
D > F	F > D	v
E > G	G > E	v
F# > A	A > F#	v
G > B♭	B♭ > G	v
A > C	C > A	4 to 6

TUNING: ADDING THE PROFESSIONAL TOUCH 169

Scale (2)

SCALE NOTES	INTERVAL	BEAT RATE PER SEC.		TEST MAJOR 3RDS	TEST MAJOR 6THS
C < G	4th	1.35	Greater than pure		
C < F	5th	2.0	Less " "		
G > D	5th	1.65	Less " "		F > D 2.5 Beats greater than pure
D < A	4th	2.25	Greater " "	F > A pure	
A > E	5th	1.9	Less " "	C > E pure	G > E 3.0 Beats greater than pure
E < B	4th	2.5	Greater " "	G > B pure	
B < F#	4th	1.9	Greater " "		
F# > C#	5th	1.5	Less " "	A > C# pure	

As we wish to leave the wolf between A♭ and C#, we must now proceed around the other side of the gamut of the scale.

F > B♭	4th	1.5	Greater than pure	B♭ > D pure	
B♭ > E♭	4th	2.35	Greater than pure	A♭ > C pure	
E♭ < A♭	5th	1.7	Less " "		

Notice at this stage, we now have six pure major test 3rds and 2 major test 6ths.

MAJOR 3RDS	RECIPROCAL UPPER MINOR 6THS	BEAT RATE PER SEC.
A > C#	C# > A	Pure
B♭ > D	D > B♭	″
C > E	E > C	″
D > F#	F# > D	″
E♭ > G	F# > D	″
F > A	G > E♭	″
G > B	A > F	″
A# > C	B > G	″
A > C#	C > A#	″
	C# > A	

We now look at the progression and balance of the minor 4ths against their reciprocal upper major 5ths.

MINOR 4THS	RECIPROCAL UPPER MAJOR 5THS	BEAT RATE PER SEC.
A > D	D > A	1.125
B > E	E > B	v
C > F	F > C	v
C# > F#	F# > C#	v
D > G	G > D	v
D# > A♭	A♭ > D#	v
E > A	A > E	v
F# > B	B > F#	v
G > C	C > G	v
A > D	D > A	v
B♭ > E♭	E♭ > B♭	2.25

Irregular meantone temperament

All tunings so far have been regular tunings. Irregular tunings, with a mixture of tempered and untempered intervals, emerged at the end of the 17th century, continuing into the 18th century, in which two 5th intervals, E♭ > B♭ and B♭ > F, were tuned greater rather than less than pure.

This irregular tuning suits the harpsichord music of Louis Couperin very well, especially the pieces with flat key signatures. To set this temperament, tune the regular meantone temperament (1) as far as G# (see page 162).

TUNING: ADDING THE PROFESSIONAL TOUCH

Extend the scale downwards and upwards for the notes tuned. We now have only the two remaining flats to set.

Balance the E♭ below middle C with the G of the bearing (major 3rd) and with the B below the bearing (major 3rd) to make two good 3rd intervals, the upper interval beating slightly faster than the lower one.

Next, balance the B♭ of the bearing with the F of the bearing (4th interval) against the E♭ already tuned (5th interval) so that it makes as good a 5th with E♭ and as good a 4th with F as this will allow, bearing in mind that the 5th must be greater than pure.

Tune the remainder of the E♭s and B♭s to complete the tuning. The wolf 5th still remains between G# and E♭ but is slightly less than before.

If we make the tests which we ran on this same scale before the two note alteration, and compare them, we will see the subtle changes which have taken place, the progressional beat rates being interrupted by odd faster beat rates.

MINOR 3RDS	RECIPROCAL UPPER MAJOR 6THS	BEAT RATE PER SEC.
A > C	C > A	2 to 3
B > D	D > B	v
C > E♭	E♭ > C	14 to 16
C# > E	E > C#	v
D > F	F > D	v
E > G	G > E	v
F# > A	A > F#	v
G > B♭	B♭ > G	8 beats
G# > B	B > G#	v
A > C	C > A	4 to 6

MAJOR 3RDS	RECIPROCAL UPPER MINOR 6THS	BEAT RATE PER SEC.
A > C#	C > A	Pure
B♭ > D	D > B♭	4 beats
C > E	E > C	Pure
D > F#	F# > D	Pure
E♭ > G	G > E♭	Rapid
E > G#	G# > E	Pure
F > A	A > F	″
G > B	B > G	″
A > C#	C# > A	″

MAJOR 3RDS	RECIPROCAL UPPER MINOR 6THS	BEAT RATE PER SEC.
A > D	D > A	1.125
B♭ > E♭	E♭ > B♭	3.0
B > E	E > B	v
C > F	F > C	v
C# > F#	F# > C#	v
D > G	G > D	v
E > G#	G# > E♭	6.0

Interval added to scale

E > A	A > E	v
F > B♭	B♭ > F	v
F# > B	B > F#	v
G > C	C > G	v
G# > F#	F# > G#	v
A > D	D > A	2.25

Temperament ordinaire

In the 18th century, by the time of D'Anglebert, a system which became known as Temperament Ordinaire was being widely used, which had eliminated the wolf from the meantone scale. Systems similar to this were well known and used as early as 1690 by Werckmeister in his Preferred Organ Tuning.

SCALE NOTES	INTERVAL	BEAT RATE PER SEC.		TEST MAJOR 3RDS	TEST MAJOR 6THS
C < G	4th	1.0 Greater than pure			
C < F	5th	0.95 Less	" "		
G > D	5th	1.25 Less	" "	F > D 4.5 beats per sec.	
D < A	4th	1.65 Greater	" "	F > A 4.0 beats per sec.	
A > E	5th	1.95 Less	" "		
E < B	4th	2.15 Greater	" "	C > E 5.0 beats per sec.	G > E 5.0 beats per sec.
B < F#	4th	0.95 Greater	" "	G > B 4.5 beats per sec.	

At this point, all of the diatonic notes (the white ones) and the F# have been tuned. Extend the scale downwards and upwards for these notes.

Balance the E♭ below middle C so that the E♭ > G (major 3rd) beats slightly faster than E♭ < B (major 3rd), making as good thirds as the system will allow.

Balance the B♭ of the bearing so that the B♭ < F (4th) and B♭ < E♭ (5th) are as pure as the system will allow.

Make C# pure with F# and G# pure with C#.

A very fine adjustment upwards must be made to the F#, just enough to make it impure with C# but not enough to make it pure with B. This now corrects the major 6th interval F# > E♭ to beat a 4.75 beats per sec.

Extend the scale downwards and upwards as before.

The 3rds and 6ths are not now pure and have started beating again, although slower than in tempered tuning. The wolf has disappeared and the cycle of 4th and 5th intervals are complete throughout the gamut of the scale.

These changes bring the scale as close as one can get to tempered tuning without actually setting it.

The bearing intervals should be approximately as follows:

MAJOR 3RDS	BEATS PER SEC.	MAJOR 6THS	BEATS PER SEC.
F > A	4.0	F > D	4.0
F# > B♭	Rapid	F# > E♭	4.5
G > B	4.0	G > E	5.0
G# > C	Rapid		
A > C#	4.5		
B♭ > D	4.5		
B > E♭	5.0		
C > E	5.0		

Extend the scale upwards and downwards in the usual way.

The uneven beating of the minor 3rds which is reflected in the 6ths because of the mixture of tempered and untempered intervals makes the progressional testing of them not worth the effort, tests being limited to 4ths and 5ths.

MINOR 4THS	RECIPROCAL UPPER MAJOR 5THS	BEATS PER SEC.
A > D	D > A	0.8
B♭ > E♭	E♭ > B♭	v
B > E	E > B	v
C > F	F > C	v
C# > F#	F# > C#	v

D > G	G > D	v
E♭ > G#	G# > E♭	v
E > A	A > E	v
F > B♭	B♭ > F	v
F# > B	B > F#	v
G > C	C > G	v
G# > C#	C# > G#	v
A > D	D > A	1.6

Nothing serves to show the small amount of change in the 4th and 5th intervals which can make such a great change in the 3rd and 6th intervals as much as the small change necessary to Temperament Ordinaire, to change it into equal temperament.

If we compare the beat rates of the 4th and 5th intervals in both systems, we find that an average change of from 0.4 to 0.62 beats per second is all that is necessary to change the major 3rds and 6ths from beating at between 4 and 5 beats per second to between 7 and 12 beats.

The increase in speed of the major 3rds and 6ths to 4 beats per second was achieved by a change of approximately 0.5 to 0.8 beats per second from the condition in Pythagorean tuning when they were pure.

It had been my endeavour to show some of the changes which have taken place in the field of tuning, ending up back, or almost back, to Tempered tuning. There are many other variations which have not been covered and whilst it has not been my intention to do so, I feel that there is a tuning which is appropriate for each of the periods in which they were in vogue.

The complete tuner, who has taken the trouble to master the art of these early forms of tuning, is in a position to offer the musician alternative solutions to the difficulties they experience in their exploration of the works of the early composers.

9

SETTING UP THE BUSINESS

Presentation

Although the main thrust of this book is directed at producing a reference for piano problems, it would be unwise to ignore the tuner's own problems.

The first thing which happens in professional life is the meeting of tuner and customer, and first impressions are of utmost importance. In days gone by, piano tuners wore a top hat and as highly respected technicians were, in company with doctors, the only members of the professions allowed in through the front door of the great houses.

The proliferation of piano makers, coupled with the associated increase in numbers of piano tuners has led to a lessening of status as pianos have become a normal item of household entertainment.

As time has gone by, many music houses have deteriorated into glorified piano supermarkets, reducing workshops to the bare minimum, subcontracting the responsibility for the future maintenance of their instruments to students fresh from college with B Tec qualifications, without providing a continuation of post-college training under experienced supervision.

This deterioration started with conditions of employment imposed by the government of the 1970s. As with so much legislation designed to protect employees, they have resulted in just the opposite effect.

The charges made for many service industries could not support the imposed legislation and the services provided have been thrown back into the lap of former employees as subcontractors.

In these cases, the firm decides on the rate for the job, plus VAT, and takes a percentage to cover clerical services, passing on the remainder to the subcontractor. This is invariably not enough to make adequate provision for all of the items which government legislation dictates that the firms should provide and sometimes results in a skimped job in an effort to gain the income which service personnel feel is their due.

However, we live in the world as it is, which is not perhaps as we would like to see it. Without proper guidance and further training under supervision, attitudes to dress and manners are left to individual taste or lack of it.

The casual approach to dress and manners of the 1960s has carried through to today; in many cases, open-necked shirt, bomber jacket, jeans and trainers seem to be the norm. The impression this gives is that the same sloppy attitude will be applied to the job in hand and this is not an auspicious start. If the student attitude to dress is carried forward into professional life, it can cause considerable problems and a considerable loss of earnings.

A case in point happened some time ago, when I was called to tune a grand piano in a first floor flat in a well-to-do local area. After the initial inspection and chat to find out what was required of the instrument, it transpired that the lady would be making only casual use of it as the mood took her but she wanted it to be kept in reasonable order. As the instrument was in quite good condition, the tuning commenced. Light refreshments duly arrived and a further conversation ensued, during which she confided in me that previously she had telephoned our large local music house for a tuner. The result was that a motor cycle pulled into the drive and, in her own words, 'A heavy-booted, long-haired, leather clad, bejeaned figure came to the door.' Her first thought was that it was one of her son's friends and when he announced that he was the piano tuner, her immediate reaction was to say, 'No! Not here, there must be some mistake,' and she shut the door.

'I was not letting him loose on my piano,' she snorted indignantly. I said, 'But he might have been a very good tuner,' but no amount of persuasion was going to allow him over that door step.

Let us look at the situation from the financial point of view. The best instruments are owned by the more affluent members of society. These in general are the ones more willing to have their instruments tuned regularly over a number of years.

By the same reasoning, they are the ones who will be unwilling to have this service performed by anyone who does not conform to their own standards of manners and dress. It is not much use being the most skilled tuner in the area if you can't get over the door step to prove it.

There are always areas of work where dress and manner do not matter much. There is the customer whose choice of tuner is governed solely by cost, and is willing to put up with anything to this end. It is quite usual for me to get calls from unknown sources making enquiries as to the cost of piano tuning. (This can come from individuals or from other firms testing the state of the market.) On being told, they sometimes express incredulity, a reaction nearly always caused by the fact that they have not called in a tuner for a very long time.

On such occasions, I always offer them the telephone numbers of tuners whose charges are much less than my own. This usually elicits the query, 'Are they any good?' My reply is always that with their lower quality of service they are adequate for less discerning musicians. This puts the callers on the spot; they have to admit that they are musicians of this class, or that they are content with the lower standard.

Such expressions as 'It's only for the children' are common, and it's no use wasting time telling enquirers that only the best is good enough in the learning stage; they don't want to know. In these cases, the talent of the best tuner is wasted.

Local authorities, schools and hospitals are areas in which cost is more important than personal presentation, often accepting only the lowest tender, with the resultant abysmal standard.

The average household, music teachers, professional musicians, clubs, music colleges and the more affluent members of society make up the other end of the scale and are better from the financial and from the quality tuning points of view. It is here that personal presentation is extremely important.

In my own case, on returning to my firm after service in the Royal Navy during the Second World War, I was given a period of improver training at the factory and set to follow an elderly tuner of seventy-five, retained past retirement due to the war. We shall call him Mr Brown.

It was prudent to find out as much about him as possible in order to decide on my own presentation. There were many occasions on which tuners could meet and chat, such as pay day and when required to work in the factory. Bus, tram, ferry and rail points were all opportunities for a chat, as in 1946 not many tuners belonged to the car-owning fraternity. It was always possible to find two or three waiting at any one of these places at any one time.

Discussions were invariably of a personal or work nature. If, however, Mr Brown arrived on the scene, this was the signal for the dispersal of all but the unwary, who were belaboured with a long tirade on the conditions he had to endure at home from his family.

This was his method of eliciting tea and sympathy from his

customers. It was not unusual for him to arrive at 8.30 in the evening to tune the piano, not leaving before 10.30. He would puff away on a pipe, leaving rooms reeking of tobacco smoke. It must be remembered that smoking at this period was the norm and because of this, plus the shortage of tuners, he was accepted. When I was set to follow him, he would tell his customers that there would be a young man coming next time, and if there were any problems, he would put it right when he called next time. Not an auspicious start for me.

From what I had learned about Mr Brown, I decided that my presentation would be very different. Being of a naturally happy disposition, a non-smoker, a pianist (which he was not), and taking trouble to dress in a more professional manner, I was going to create quite the opposite impression. Although some customers still wanted me to spend Mr Brown's customary two hours tuning their pianos, by making sure that there were no causes for complaint, by playing the instrument and inviting comment on the finished job.

I succeeded in winning over ninety-five per cent of his customers, as he gradually relinquished first the outlying, and then the more central area tunings. As the number of customers asking for my services increased, so did my tuning round. The work of Mr Brown alone was not enough to keep me going, and other tuners were asked to pass on work from their round and have it replaced by new work. Human nature being what it is, the work they gave me was the worst and most difficult, or the customer was particularly difficult to deal with.

This was how I came to have an unusually large number of difficult models and customers on my round, and later on became the firm's troubleshooter. In tuning miniature pianos of all types, I must have spent more time on my knees than the vicar of the local church.

Customer relations

Dealing with people comes easily to me, due in the main to a naturally happy disposition. To make the wrong impression from the start can be catastrophic. We have discussed a clean appearance. One does not have to look as if one has just stepped out of Burton's window to create this.

A smile is the most disarming of first impressions. A polite and respectful attitude, even if this meets with a stony stare, will eventually break down the natural reserve of the most difficult nature. One does not have to adopt an obsequious attitude or act in a way which is unnatural to one's own personality.

'Chatting up' is the worst first impression of all as it at once puts the wary on guard. Conversation should be limited to the job in hand and should develop as a natural consequence of that.

If you smoke, don't do so until invited to, don't even ask, as, although the customer might not like to refuse you, they may not send for you again. On more than one occasion, a customer has complained that following a tuner's visit a burn mark has been found on a key that was not there before his visit. If the tuner had been smoking, he is left with no proof that it was not his fault. It may well be a 'try on', but it is very difficult to prove and if nothing is done about it, customer relations will suffer.

On completion of a recent routine tuning, a customer called attention to a tiny piece of veneer missing from the case, assuring me that it had been all right before my visit. Prepared for such eventualities, I carry small pieces of veneer and a touch-out kit. The size of the piece was too microscopic to replace, and the space was matched for colour with a wax mixture and filled to her satisfaction. The size of the veneer that was missing leaves me with the impression that she knew of this before my visit and that it was a 'try on'. As I was prepared for this and it only took minutes to disguise the space, by my accepting the situation as it was, customer relations remained good.

On my next visit, before tuning commences I shall make a minute inspection of the instrument for scratches and veneer splits, drawing her attention to any which do exist. This will have the effect of making her aware that I am watching for any further attempt in the same direction.

I have mentioned that in the past it was usual for tuners to travel by public transport, and in conditions of deep snow I would arrive in Wellington boots but would also have a pair of light-weight slippers to change into at the door before venturing into the house and walking across the Persian carpet. Comments such as 'That's very considerate of you, thank you!' immediately establish an excellent customer relationship. To clomp across the same carpet in heavy motorcycle boots does not give the same impression.

This sort of consideration conveys, without spoken words, *professionalism*.

Do not however, run away with the idea that I never have problems. Some case histories may serve to illustrate this.

It is possible to make innocent mistakes which, if care is taken can be avoided. Calling at a large house situated in its own grounds with large imposing gates and a long drive, I eventually arrived at the house. The door was opened by a young man of about thirty-five.

'Mr Harvardson?' I enquired. 'It's the piano tuner.' He smiled, inviting me in, taking me through to a large room and introducing me to Mrs Harvardson, an elderly lady of about seventy, heavily made up in a garish clown-like fashion, dead white face, a circle of red on each cheek and a smudge of red lipstick in the centre of her large mouth. She was dressed in black, necklaces were draped around her neck and there were rings on every finger. Leaning heavily on a stick, she led me into a second room which contained a large Bechstein grand in a considerable state of wear. In conversation, it transpired that she had been a music teacher of some considerable reputation.

Tuning completed, she returned, removed the multitude of rings and commenced to play. Obviously, in spite of her age, she was still an excellent pianist. Expressing her satisfaction, she said, 'My husband will be pleased now. I'm teaching him to play you know.' Commenting that this was most commendable, I asked if her son played. 'I have no son,' she replied. 'Oh!, I thought the gentleman who let me in was your son.' There was a pause, 'That is my husband,' she replied coldly. The moral of this incident is, *never assume anything*.

Since that incident, I have cultivated the habit of referring to other members of the household as 'the lady' or 'the gentleman' until the relationship is revealed.

I have been called to many instruments which have been cleaned thoroughly by the housewife to the extent that I could not have done better myself. However, this gallant band of ladies has long since disappeared, and it is customary to ask for a duster if one has not been left on the piano to reinforce one's own dusting brush to clean out the more inaccessible parts. Even this innocent request can create the wrong impression. I remember arriving at a large terraced house in Liverpool, the outside of which fairly shone, crisp bright paintwork, neat curtained windows, the steps outside and even the pavement well scrubbed. The door was opened just enough to disclose an eye.

'What is it?' said the voice in a strong Scottish accent. When I established the reason for my call, the door was opened wider to disclose a thin, grey-haired lady in a drab black dress several sizes too large for her. Her thin, heavily lined face gave the impression of having lost a great deal of weight.

'Wipe y'rr feet' she demanded. Then, reaching down behind the door, she produced a large bundle of newspapers and commenced to lay them down along the hall over the canvas overlay already in place for the carpet's protection. Following her as she backed along the hall, I eventually reached the front room door. She opened it, revealing a room full of heavily shrouded furniture. A canvas cover completely covered the carpet, on top of which she began to lay more newspapers.

Raising a cover from what was obviously by its shape the piano, I met a large black Eberhardt upright of First World War vintage, complete with candle holders polished until they shone like gold. On removing the top door, and laying it down on the newspapers, I made my first mistake: I asked for a duster.

'Duster? Duster?' she fairly screeched. 'Y'll find nae dust in here, young man.' Apologising for my stupidity, I explained that what was a normal request in other people's homes was obviously not required here; not everyone kept their homes as spotless as this. She beamed with obvious pleasure. But her face darkened again. 'Y'll nae smoke in the hoose.' It was a demand rather than a request.

This is where I made my second mistake. I assured her that I did not smoke at all, the smile returned and I could see that I would be saddled with this call in the future.

The next call was only two houses away, always a bonus when travelling by public transport. This was a different household altogether. A large jolly woman opened the door. 'What has happened to Mr Meek?' she enquired. I explained that I had recently returned from the Forces and Mr Meek had kindly donated some of his calls to me.

She roared with laughter. 'He's terrified of Miss McGraw,' she laughed. 'That is why he gave them to you. You'll be in need of a cup of tea,' and she left me to the tuning.

Many years later, my own apprentice Jeremy Stubbings attended Stephenson's College in Edinburgh. This college was chosen for him as they allow students to spend one term in each academic year back with their sponsor to do the practical work and cover the area of industrial experience lacking in most colleges. During these periods, Jeremy made notes of all the diagnoses and problems encountered in practical training, putting this forward at the end of his training as a final document of his accumulated experiences. Jeremy asked me as his sponsor to write an introduction to his efforts and to this end, I penned the following.

ODE TO JEREMY

When you were young, the task was set

For you to learn your alphabet.
First there's 'O's and then there's 'A's,
Who said these were your happiest days?

At last decisions have been made
To gain a skill and learn a trade.
Now it comes and this makes sense,
Industrial experience.

Each day you go out with 'The Boss',
To try your hand what e're the cost.
Replace a flange or fit a string,
Whatever each day's work will bring.

Our first call is on Mrs Brown
Resplendent in red dressing-gown
Opened far beyond the knee
Posing there for all to see.

Jeremy's face is very red
'She looks as if she's just from bed.'
And his eyes grow wider still,
As she leans forth to pay the bill.

Our next call is on Miss McGraw,
Of craggy face and granite jaw.
Thin of shoulder, slim of hips,
The words just slip between tight lips.

The door is opened just a slit,
She eyes us saying, 'What is it?'
We state the object of our call,
'It's in the room just down the hall.'

The room pervades an air of gloom,
White dust sheets shroud the whole room,
We raise a cover and we find,
Piano of the older kind,

Candlesticks that gleam like gold,
Inlay and the name enscrolled.
To ask for duster would be folly,
We'll find no dust in here, by golly.

SETTING UP THE BUSINESS

Down the avenue we call
On jolly Mr Bernard Small,
Six feet two and nineteen stone
Loud voice and to laugher prone.

He fills the door from wall to wall,
We squeeze on past him down the hall.
'It's in the back room, boys,' he bellows,
We ease our way past bass and cellos.

Minipiano we trace
'Twixt window and the fire place.
Job completed, asked to test
Mr Small performs his best.

Gigantic man, piano small,
Rendering of Elija's Call.
Satisfaction guaranteed
We take the money and proceed.

Mrs Jones is small – petite
Tiny hands and tiny feet.
Very proud of the Welsh nation
Suffering – she is – from vibration.

Rumbling it seems to be
Very bad on middle D.
We listen hard to trace direction,
Very bad, this middle section.

Searching, find a 'Ping-Pong' ball
Down between the back and wall.
'Duw! those kids are very bad,
This will make my 'usband mad.'

Now it's time for lunch and talk.
We stretch our legs and take a walk,
Eat our sandwiches by the river,
Paté paste of ham and liver.

The afternoon brings Ada Phipps,
Musician to her finger tips.
Housework, though, is very rare,
Dirt and cobwebs everywhere.

Cats abound in twos and threes
Dodging swiftly 'tween our knees.
Glaring at us balefully,
Top of sideboard – chair – settee.

Ginger tom sits on the mat
Eats a tin of Kit-E-Kat
Jeremy's face a sight to see
As he says quite close to me.

Sweeping music to a chair
Thus revealing Steinway there.
Cats have been in here as well,
Leaving truly awful smell.

Cats collect before the fire
As we adjust a damper wire.
Tuning finished very quickly,
Leaving feeling rather sickly.

'Come upstairs and tune the upright.'
This makes us feel rather 'Up-tight'.
Clambering up the wooden treads
Noise vibrates within our heads.

Doorways 'click' as we pass by
And at gap we see an eye.
Doors that open just a crack
Close again as we look back.

Start on upright – awful sound,
Sound that echos round and round.
Doorway opens – man appears,
Trilby hat, a scarf and beard.

Glares at us. 'Will this not cease?'
Screams, 'Oh God!, Is there no peace?'
Goes inside and slams the door,
Hoping we'll come back no more.

Job completed, homeward bound
Head still ringing from the sound,
Nostrils reeking from the smell,
Next time she can go to hell.

When I die, as die I must,
I must leave someone I trust,
To tend my business, and my friends,
When the reaper's scythe descends.

Later on, as time goes by,
And you are as old as I,
One day when the going's tough,
And you feel 'That's near enough',

From a place far, far away
You will hear a faint voice say,
As ghostly hand lays on your scruff
'Near enough's not good enough.'

But just a minute. I'm not dead
There's lots and lots of time ahead,
So tend the bearings, get them straight
Count the thirds that beat at eight.

Back to college you must go,
Practice tuning, make it flow.
Concentrate and persevere
And I'll see you again next year.

Although every day's work is not like this, thank God!, every incident described in the Ode has happened to me. Learning to cope with situations like this is one of the distinctly necessary qualities to be learned by experience. All that a book can do is to point the way.

It may seem strange to stress such things as wiping one's feet on the mat. Not to mention closing garden gates both on arrival and departure. One only has to experience a customer in curlers and dressing-gown charging off down the road at nine o'clock in the morning after an escaped prize bitch on heat and pleasure bent. More serious is the risk of children escaping onto a dangerous road. Just because the gate is open on arrival is no reason for assuming that this is usual. If the gate will not close, then it is safe to assume that this is normal for the household.

Dogs can sometimes be a trial, if precautions are not taken in advance and preventive measures adopted. To retain a professional appearance, I wear washable suits which I can change and press regularly. On changing my suit, I make an effort to make friends with two or three friendly male dogs, allowing them to rub against my trouser leg as I pat them. This is a certain way of gaining an interesting scent for subsequent dogs (also preventing them from looking upon my leg as their next meal). A growling or barking dog will suddenly stop on smelling the scent of other dogs and will soon lose interest when the smell poses no threat to their domain. Do not allow a bitch on heat the same privilege or one's leg might receive attention of a more loving kind.

Children can sometimes be a problem in the presence of such a delicate mechanism as a piano action, and a grab at a moving part which is a wonderland of magic to them can prove disastrous. A firm but friendly hand and a wary eye is usually enough to deter them, but if this does not work they must be removed by the parent in the interests of safety of both piano and infant whilst you explain that sharp bridle wires can injure little fingers as well as being expensive to repair.

Cats can also be a problem due to their enquiring nature and must be watched if they become too inquisitive.

Whilst having a problem closing a bottom door due to the fact that the case had swollen with damp, my attention was directed to fitting the bass end. Without my knowledge, the cat walked into the piano at the other end. Finally replacing the door, I called the customer to say that I had finished and as I was making out the record card, a sound came from the piano. We looked at each other. 'It came from the piano,' she said accusingly. I agreed, walked across to the instrument and commenced to play it, still unaware of what had happened.

A loud mewing accompanied by a violent scratching came from inside the piano, I hastily removed the bottom door, out shot the cat, heading for the Argentine judging by the speed, pursued at the same speed by the owner. She returned a few minutes later breathlessly clutching Tiddles to her ample bosom. She thrust the money at me, I hastily gave her a receipt and an apology, and made a hurried exit.

Dusters come in many forms, from the paper type 'J' cloths through the range of yellow dusters, pyjamas, nightdresses to the daintiest of undergarments. Only a few weeks ago, I was handed an extremely scanty pair of knickers to use as a duster, with the smiling remark, 'I'm sorry but it's all I could find.'

'That's all right, Mrs J,' I replied. 'As long as they're not still warm, I don't mind.' She left, laughing uproariously.

One must never read into a situation like this anything more than a rather slapdash attitude to housework. If, however, I have ever detected any attitude developing beyond the strictly business relationship, I have found that by bringing my wife and family into the conversation and in turn asking about the lady's family seems to solve the problem.

We all have attitudes of dress and manners which we project outside the home while adopting an entirely different attitude when relaxing within the confines of our own home. To spend the time it takes to tune the piano within the confines of someone else's home can be an eye-opener and must be coped with without the slightest sign that what occurs is in any way unusual.

One of the most frequent and difficult problems is the situation when the tuner finds himself between the parent and pupil pianist. A frequent scenario is that the tuner inspects the instrument, asks who plays and what standard of proficiency has been attained. This will establish the playing-to-tuning ratio required and the probable demands both present and future that will be asked of the instrument. With this knowledge, sound advice can be given as to present usage and future potential.

Some years ago, I was called to a wooden framed piano *circa* 1860 of an indeterminate make. Although everything worked, my enquiries revealed tht the owner's daughter was taking the Associated Board's Grade 4 examination and was complaining that it was out of tune. It was made quite clear to the customer that an instrument of this poor standard was unsuitable for this or any other grade. The owner insisted however that her daughter was only complaining that it was out of tune and as long as this was put right that was all that was required. Reluctantly I tuned it.

Several days later, I received a complaint from the same house that the piano was still out of tune. In all cases of this sort, I make arrangements to call when I can meet the pianist. On arrival, I was met by a sixteen-year-old young lady who, when asked to demonstrate where the instrument was out of tune, related problems of touch and tone that were obviously absent. Tuning as such was never mentioned. Mother was brought in and made to confirm in front of her daughter that she was already aware before tuning commenced that the instrument was unsuitable for even grade 1 let alone grade 4.

The problem was then left to be settled between mother and daughter. This is a typical case of the tuner being used by two people each treating the word 'tuning' as if it were a magic wand or a magic lamp which can cure everything, turning old pianos into new. We are ordinary mortals, working within the scope of the instruments presented to us for service, and should not be used as a buffer between pianist and the parents' inability to provide an adequate instrument.

We expect top-class musicians to appreciate our problems and limitations, but this is not always so. In musical terms, pitch has always been in the process of being raised and has resulted in tenors and sopranos of long ago being accredited with the ability to reach higher notes with greater ease than today, and is also the reason for modern tenors and sopranos burning out early in their careers.

The Royal Liverpool Philharmonic Orchestra uses the pitch of A441, which has been explained by the leader as the sound he expects to hear when the orchestra plays. As far as the Philharmonic Hall is concerned, this presents no problem to the tuner. It is only when an attempt is made to impose this standard elsewhere that problems occur.

One morning I received a call from the theatre manager of the Floral Pavilion in New Brighton, advising me that a jazz group from the Liverpool Philharmonic would be performing on Saturday and had requested that the piano pitch be raised to A441 for this performance. It was explained to him that in order to do this, I would have to disrupt the stability I had so carefully cultivated over recent years and also return it to A440 again after the performance.

When advised of the charge I would make, plus the fact that I would not guarantee for some time afterwards the stability of the instrument for equally celebrated musicians, he advised them that until the recognised standard of A440 was raised nationally, the standard at the Floral Pavilion would remain at A440.

Music societies can sometimes place tuners in considerable difficulties. Having invited Peter Donohoe to perform at their town hall, one of my local music societies refused to pay the cost of hire and removal of the local authority's Steinway, substituting a grand piano from the front parlour of a society member.

Arriving to tune the piano prior to the afternoon rehearsal, I was horrified to see a 5' (152 cm) Danemann grand in very poor tuning condition. The secretary assured me that the owner had the instrument regularly tuned by Harrods. Someone was obviously lying. After protesting at the inadequate instrument provided and the poor state of the tuning, I insisted that the secretary stayed to meet the artist, offering him any explanation necessary.

In order to make some headway, I began the tuning, which took two and a half hours, finishing just as the artist arrived. A consultation on the inadequacies of the instrument ensued, and with the proviso that the instrument would be retuned after rehearsal, plus a retune during the interval, for which I was granted thirty minutes, the concert proceeded. The cost of removal and return after the concert, the charges for all this attention plus the cost of another tuning at the house after the return, made precious little saving on the normal hire charges of the authority's Steinway which would have provided a far superior concert.

This is a situation I have taken great pains to avoid since that date, making certain that the instrument has been agreed with the pianist before accepting the commission to tune.

Most internationally known artists do check the type of instrument they will be required to play, and on arrival have a detailed knowledge of its quality, even modifying their programme to suit.

In forty years of this type of work, I have never met from a pianist of quality any request for adjustment of tuning or regulating which was not justified. This may be because, being a pianist myself, the pianistic problems described to me in pianistic terms are easier for me to translate into tuning and regulating terms.

One such occasion occurred after an afternoon rehearsal, when the artist complained that the pedals were not working effectively. On examination, they were moving down only about half an inch (12 cm) and the dampers were not clearing the strings. A check to see if the lyre had slipped showed that it was quite secure and there was no delay in the lift, therefore no adjustment to increase it. Noticing that the pedals were actually touching the stage when fully depressed, and being underneath the piano at the time, I noticed that the treble leg had disappeared through the stage up to the hub of the castor. Hastily, I got out from underneath and enlisted the aid of the stage hands to lift it out to a more secure position.

An extension had been erected for the accommodation of a large choir, and was not constructed to accommodate the weight of a grand. A discussion with the manager produced three metal plates to go under the castors to distribute the weight, and three shallow castor cups to ensure stability. All problems are not of a purely technical or pianistic nature.

At another concert at a local public school using their own Steinway grand, the pianist complained in the interval that the G above middle C was missing occasionally. Removing the action, the offending note was adjusted with no apparent effect. A closer inspection revealed a broken wing spring. In the time available, there was nothing for it but to move a section from the far treble to the middle. After a consultation with the pianist, to find a note that would cause him no problem, the sections were changed, regulated, and the concert proceeded, the spring being changed afterwards.

To cope with problems in such circumstances strengthens customer relations and brings considerable prestige.

There is always the customer who cannot be satisfied no matter how hard we try. Such a person was Myra Brass, violinist, cellist and excellent musician. There was no way that she could be made to understand that the intervals in the bass section of her piano could not be made to produce the complete range of progressions she could produce on her cello.

When the interval changes she suggested were applied and the interval distortion these produced pointed out to her, she kept saying, 'But it shouldn't.' No explanation of the problems inherent in pianos would satisfy her and as far as I know, she is still searching for the mythical perfection she desires.

Don't expect one hundred per cent success, but keep trying.

Range of tools carried

In days gone by, when tuners travelled by public transport, it was not possible to carry a large selection of tools but as the cost of labour was low, it was possible to send someone from the factory at a reasonable charge once the tuner had established the type of job and the tools required.

This is not the case today, but as tuners in general have their own transport the range of tools can be greatly extended and the range of household services increased. In my own case, on looking around at the services offered by my competitors both large and small, there was obviously a gap in the market for the instant 'on site' service of all but the major overhaul. Items such as the instant repairs to grand legs, fitting of castors, safety brackets, slings, footboard and casework repairs, replacement of pedals slung underneath the footboard, etc., could all be accomplished by having the right equipment available.

The Broadwood bridge repair on pages 28–30 was completed in the front room of the owner's house, not in my workshop, the obvious saving in transport costs being passed on to the customer. It does not take long for word to get around that having received an astronomic estimate from a major music house the same job was completed at half the cost and without the removal of the instrument.

This can only be done where there are no children to interfere with the instrument while it is resting on the bench trolley. School piano castors which are unsafe, or are digging into the caretaker's carefully polished floor, can be instantly replaced, requiring only a phone call to the tuner. The school piano with a collapsed foot board could be refitted instantly, all at a profitable charge which none of my competitors could match.

Castors and grand trolley frames

Take care in the use of a folding bench trolley. Check the wheel clips for security, they can drop off during transport. Find the most secure place to fit a grand lifter, making sure that everything and everybody is kept clear during any of the subsequent operations. Do not accept help from anyone. They are not insured, YOU MUST BE.

In schools and public places, the safety of pianos is essential and this must be reflected in the attention given to them. The castors must be of sufficient strength to support the weight of the instrument and move it with safety.

As the greatest weight is in the metal frame, to the rear of an upright piano, there is a tendency for it to topple over backwards, when the rear castor is fitted directly beneath the frame. By extending the castor to the rear by means of an extension bracket, this problem is greatly reduced, and is essential in conditions where it is to be moved frequently.

Shepherd (ball type) castors are to be avoided, as the shank tends to give way under the weight of most instruments and the ease of movement gives it sufficient impetus, once on the move, to overcome the best extension brackets, with disastrous results.

Twin-wheel rubber-tyred castors and metal safety brackets are available and suitable for the largest uprights. For the smaller instruments, the metal sling with its associated castors are ideal.

There are instruments with the castors set into the base board, and in replacing these due account must be made of the level above the floor to which the replacement castors leave them. This may place the pedals at an unacceptably high level or, in some cases, leave the instrument at a higher level at the back than at the front. Quite often, a cutaway section will have to be made in either the footboard or the toe to level the instrument at the correct height. Obviously, on occasions such as this, the necessary tools will be required.

Modifications sometimes have to be made to give the safety brackets something to screw into, each one being judged on its particular construction.

Pianos supplied with wooden safety brackets fitted from new, get damaged in the course of use, and have to be replaced with metal brackets or a sling. If the right-angled wooden pieces are saved, they can be used to modify instruments with little or nothing to secure brackets to.

There is a very real danger of grand piano legs collapsing during removal. I have seen the beam supporting treble and bass legs torn completely away from a grand, leaving it flat on the floor. During removal in a school, the danger to the pupils must never be underestimated.

The leg most prone to damage is the rear single leg. If this hits a hollow in the floor as it is being pushed from the keyboard end, it invariably suffers damage. The answer is to fit a grand trolley frame, securing the three legs together into a secure fitment, there then being no single leg receiving any additional strain. The method of fitting, if followed in consecutive order, is quite simple.

- Unpack the trolley, removing all securing tapes.
- Slacken off all clamps.
- Place the trolley next to one of the legs, marking on the leg the level at which the leg meets the base of the clamp. This will give the amount which must be removed from or added to the leg to maintain the height of the pedal lyre above the floor.

   Some of the old round externally fitted cup-type grand castors require the fitting of a block to maintain the lyre height and some of the inset castors require the removal of a few inches from each leg. Hence the marking of the legs before fitting commences, thus avoiding starting a job and finding that the necessary tools are not carried.
- Mark each leg.
- Spread the trolley so that the treble and bass extensions lie outside the instrument, the rear clamp being placed close to the rear leg.
- Find a secure place to fit the grand lifter so that when the leg is removed, and the trolley clamp is moved forward, there is no risk of the tripod legs of the lifter impeding either the clamp or the replacement of the leg.
- Remove the leg. Remove the old castor.
- Remove from or add to the leg dimensions.
- Replace leg.
- Ensure that the clamping bolt does not stick out sideways. Lower into the leg clamp and tighten up.

REMEMBER that the clamp tightens up in an anticlockwise direction.

Move the grand lifter to beneath the keyboard at either bass or treble end, fitting it in such a position that it will not interfere with the fitting of the leg or the trolley, ensuring that the clamping bolt is not sticking out sideways or forwards where it can catch on the ankle of the unwary. As far as possible, make sure that it is adjustable from inside the trolley frame.

Before securing the third leg, connect the crossbar behind the pedal lyre, bringing it as close as possible to the rear of the lyre. If this is not done at this stage, it is impossible to connect it afterwards.

Position the clamps along each tubular bar in such a way that they will close the split sections at the ends of the tubes and clamp up tightly. It may be necessary to insulate the crossbar behind the lyre with a piece of felt to support the lyre from any backward movement by a heavy-footed pianist.

There are three sizes of trolley frame; standard, extra large, and twin-leg model. In practice, I have found that the standard model is at the full extent of its adjustment at 8'6" (260 cm) and prefer the extra large model at or around this length. In schools, children tend to stand on the tubular bar, causing it to bend if there is not sufficient overlap of the clamping area.

Grands of 4'6" (137.5 cm) or less often require the removal of some of the tubular steel, both inner and outer from the short end (bass to rear leg) and it is necessary to make another clamping slot in the outer tube, so be prepared with a drill and hacksaw.

All of the above requires an extensive range of tools and equipment to be carried, the most useful being the collapsible bench trolley, grand lifter, chisels, large screwdrivers, hammer, electric drill and drills, extension lead, a selection of saws, screws of all sizes, replacement castors, slings and a multitude of spare parts and strings.

As a particular piece of equipment made itself evident, or a particular part a requirement, it was added to the inventory of my van, until nearly every job could be completed.

The cost of the initial outlay was large but the rewards great. To give a list of the items required would only repeat the items in any part supplier's catalogue, the need for a particular piece of equipment being the best reason for purchase. The only limitation is the ability of the tuner/technician to perform the tasks involved.

Tuning charges

Students are advised of the charges they should make. On leaving college, they find that although major music houses are making these charges, they can't get work at these rates, due to competitive undercutting by other tuners.

This cut-throat competition can and does lead to an associated poor quality of service, extra charges having to be made for the smallest of repairs. This results in, what is to most customers an unacceptable increase in expense for which they are unprepared. Many tuners rely on this method to boost their charges to more than the top grade tuners.

By taking the rate of the best music house, deducting VAT that they will obviously have to charge, which you will not, will give a charge that can be made on which to base a quality service.

Calls to service are in ninety per cent of cases prompted by the need for a small repair, a note which does not work, a broken spring or tape, or even a small regulation.

Never underestimate the prestige of a reasonable but high tuning charge which takes into account the small repair which was the original reason for the call and is completed at no extra charge. There is always the feeling that the repair was free, when the tuner can say that there will be no extra charge other than the original tuning charge, of which they were advised and for which they are obviously prepared.

Based on this rate, reductions can be offered for subsequent regular tuning arrangements. A reduction of ten per cent will bring in work which is still at a rate higher than the cut-throat competition and at a rate lower than the best local music houses whose overheads will be greater than yours, and provide a reasonable income.

This has always been my method, until I also reached the VAT range, by which time my organisation was firmly established.

Music teachers and professional musicians often ask for discounts on this reduced charge, but personally I have only offered a discount on these charges based on the number of instruments, three being the minimum, in such establishements as theatres, schools, institutions, music colleges, etc. Professional tunings such as the local authority Steinway, because of the specialist nature of the work, have always been made at the top rate.

Contracts for local authority schools are usually by tender, some accepting none but the lowest, with the associated abysmal standard.

On three occasions I have been approached to report on this standard and on each occasion have declined unless I was allowed to comment on the amount charged by the unknown tuner, which is the critical point in all reports such as this. The authority accepting a tender which is obviously too low in comparison to the average charges being made locally, is just as guilty as the tuner who tenders at such a rate.

I have twice accepted school contracts on my own conditions, which would allow all instruments to be graded into categories, those which could be brought up to standard, those which required X number of pounds spent on them to bring them up to the same standard, and those fit only for disposal. This put the authorities in possession of all relevant facts to formulate a coherent policy for the future.

It was not unusual on taking on a contract such as this to find that on entering a school with six pianos all were on differing pitches and the teachers expecting the job to be completed in twenty minutes, 'Just like the last man'. Gradually, as the improvement became apparent, they were the first to notice that they could use all instruments with equal facility and without problems. This appreciation brought in a considerable amount of work from the teachers themselves. Also a great deal of work was generated for my growing workshop, by way of action repairs, fitting of castors and grand trolleys, etc.

These authorities stayed with me for twenty years, until my retirement, when they were handed over to my former apprentice Jeremy Stubbings.

All authorities are not so prudent as this and reach the stage where they receive no replies to invitations to tender in spite of circularising all advertising tuners and music houses, and are placed in the position of writing to the tuner of a neighbouring authority.

On receiving such a letter, I asked to be allowed to examine the instruments involved and decided, as it was not a large authority, to accept the offer on terms that would allow the same service to be performed that my current authorities enjoyed. This lasted only for four years, when, on not receiving the contract for the fifth year, the downward spiral began again. Some six years later, the same approach was made, but the offer was declined.

Area of operation

From the point of view of the area in which operations are intended, charges must take into account time and distance between tuning. This can very greatly from area to area. My own part of the Wirral peninsula is a compact region of Industrial, seaside and close residential areas all within twelve miles from my home.

The area is rich in music, having many music societies, choral societies, amateur operatic societies, a grand opera company and the Royal Liverpool Philharmonic Orchestra and Choir, many members of which live in my area. By having the local authority schools, halls and theatres, travelling time is a minimal consideration but at the same time, must not be overlooked.

Even on the Wirral, there is one area to which I do not travel, as musicians are as scarce as snow in July. Significantly, this is the same area whose unreliable attitude to quality tuning has resulted in the schools having to find their own tuners, some good, some bad. Country areas such as North Wales are very scattered and the cost of tuning is outside the range of the average musician. It is only possible to service an area such as this by living centrally in known areas containing the main sources of work. The cost of tuning will still remain high, the greater distances reducing the number of calls which can be made, this reduction being reflected in the higher cost.

Cost of operation

Employed persons tend to get paid from the time they clock on until the time they clock off, regardless of the amount of work available, accepting sick pay and holiday pay as normal. The civil servant who has a non-contributary, inflation proof pension or the worker who only looks at his take-home pay as a measure of what he/she earns completely ignoring 'perks' is sometimes guilty of measuring the cost of tuning services by the hour, considering this as comparable to their take-home pay. This is not so.

If we take out travelling time, according to area, we end up with four or five hours of actual working time. If one is a 10–to-4 tuner, even less. The 9–to-6 tuner will obviously earn more.

The cost of transport will vary with the area covered, as will the type of transport chosen. Most tuners choose a car as being the most suitable dual purpose vehicle for business and private use. For a high mileage, a diesel van would give the best performance over the longest period of time. It may not portray the same image as a sports car, but it is more likely to carry a grand action and keys home for repair.

In my own case, I started with a second-hand van, running it into the ground in an effort to save the initial deposit on a new car, from then on saving a set amount monthly towards its replacement, eventually purchasing a new van for business use only as the business prospered and the amount of repair work dictated.

Many tuners are content to limit their activities to tuning and the associated smaller repairs, and a reasonable living can be made in this way. If, however, reconditioning and major overhauls are envisaged, a place to work must be provided; the kitchen table does not last for long. My first work place was the back bedroom, until pushed out by my growing family into a purpose-built centrally heated workshop.

Quite often, the quantity of repair work dictates the need of suitable premises, and if this is so, then the repairs will in due course pay for them. Before any take-home pay is calculated, the cost of providing these items must be subtracted as must be the following.

Loss of earnings insurance
The DSS sick pay given to the self-employed does not in any way compensate for the loss of earnings suffered, this must be covered by insurance.

Insurance for the above, comes in two main types and an understanding of how they work is essential, in order to take out the correct cover in the correct amount on the right policy.

The most expensive policy undertakes to cover loss of earnings up to seventy-five per cent until retirement. An element can also be included to keep pace with inflation. The policy requires that the candidate be subjected to a medical, so the younger and healthier one is at the outset, the better.

Under this system, the assurance company can't wriggle out of its obligations if the cause of incapacity were to last for many years. They can however, impose certain conditions, such as a waiting period, from one to three months before payment commences, thus relieving them from payments for short-term incapacities. If payment is claimed under this policy, they can refuse to accept any future increase in payment to take inflation in to account.

To the self-employed, the period of delay in the implementation of the above policy can be a financial problem. This can be covered by taking out short-term insurance to cover this period. These policies are of twelve months' duration and for the same amount of income cover are considerably cheaper.

If, however, a claim is made of long duration under the policy, the company will of course pay up, but is within its rights to refuse to renew the policy for succeeding years and impose conditions excluding anything which can be faintly associated with the original claim.

Any attempt to obtain cover elsewhere must disclose any previous refusal to cover and why.

Medical cover
The best and most considerate of customers can reach the stage when they can wait no longer for one's services, and to find that

the incapacity suffered has a normal delay of X months before a specialist physician can be consulted, plus a further delay before a consultant surgeon can be see, coupled with a long convalescence, can ruin a business built up over many years.

Although life-threatening incapacities are dealt with by the NHS very quickly, it is in general the non-life threatening operations which have the longest waiting lists.

BUPA, PPP, or other companies will cover this, but make sure that there is no escape clause allowing the company to refuse further cover in the event of a claim or to make an increase in the cost which makes further cover prohibitive. Once again, the earlier this is taken on, the better.

Retirement pension payments
Too many self-employed persons do not consider retirement pension payments until it is too late, putting it off until a later date which never materialises. Those who do sometimes fail to keep payments in pace with inflation, by annual increases.

In my younger days I failed in this way, and what seemed adequate at commencement failed miserably twenty years later, resulting in massive increases in payments over the next twenty years to obtain a reasonable level of pension.

Public and products liability insurance
Not only is it illegal to operate without this cover, the financial implications should a claim be made against the tuner by way of damage to an instrument or to premises he is attending can be very great. The cost of such cover for a business such as ours is minimal, working out at less than 1p per tuning.

Telephone and answerphone
The telephone is one of the most important elements in the piano tuning profession. It is the means of customer contact, and also the means by which they contact us. It is, in fact, a lifeline.

There is nothing to match the personal touch of a telephone answered by a voice which can chat to the caller and which has a complete grasp of the overall operation, but tuners working alone must provide an answering machine of some sort, to take such things as cancellations and enquiries. Many customers find talking to a gadget which cannot answer their queries disconcerting, but if this is the only way that the contact can be maintained, it must be made as pleasant as possible.

Many answering machines enable contact to be maintained from any part of the country, by telephoning one's own home and activating the machine to see if there are any messages.

Advertising
Never underestimate the power of advertising.

This slogan has been propounded by newspapers for years, and it is a statement of truth. We all need it and it pays. The problem is to tie in the advertising with the business in such a way that the two agree. As the telephone is the main form of customer contact and the way they immediately think of contacting you, to tie the advertising in with their thoughts is good business.

First, consider the area in which operations are envisaged and which one can cover economically. Then, look for the periodical or local newspaper, free press, Thompson's local directory, Yellow Pages, etc, which cover the area most effectively. In my own case, after taking into account cost, area covered and the number of households with the particular publication in their possession, without the shadow of doubt, Yellow Pages won on all counts. The association of the telephone and the publication, plus the publication's own television advertising clinched it for me. To strengthen the association, I gave receipts on cards of the same colour yellow and later having a van sprayed to match.

If advertising of this nature is envisaged, it needs to be arranged more than twelve months in advance of the start-up date. Before the publication date, the advertisement must be designed, an eye-catching logo made which can be reproduced on a card given as a receipt which will often be kept by the customer and in many cases handed on to friends. In my own case, the logo was reproduced on my van. If a car owner, magnetic panels can be obtained which can be easily attached and detached when not required.

Look at the size of similar advertisements in the same publication and make an insertion of at least equal or greater than your contemporaries which, by inference, indicates that you are at least of the same status or greater. Many calls are prompted by the physical size of an advertisement.

Secretarial time
Of all the methods of contacting and advising customers of an intended visit, the most effective method has proved to be by letter, or post card. Although this is by no means perfect, it has the advantage of providing a means of permanent reminder to members of the household, even if their habit is to prop the card up in front of the clock until the appointed day. Such wording as

Dear Mrs Jones,
I hope to have the pleasure of calling on Thursday morning 22nd March 1991
Your tuning charge will be £X.
Yours faithfully,

is quite sufficient to let them know that you have organised a day's work in the area and that you will call unless they let you know anything to the contrary. The majority will let the visit proceed unless there is any previous engagement which prevents this.

The secret is to find the length of time which is not too far ahead and not too short notice. The postal service has varied from the predictable to the impossible. In the last few years, things have improved, to the extent that a system is possible. In general, I have found that a clear week's notice has proved the most successful, providing that the day and date is underlined in red.

Gaps in the day's work result due to an inconvenient choice of date and are distinctly necessary to allow the insertion of casual calls to service which, if advertisements are conducted successfully, will come in.

The sorting of file cards, book-keeping and the sending of these cards or letters takes up a considerable amount of time but, if good book-keeping is arranged, will keep to a minimum the cost of employing an accountant.

Accountancy

To employ an accountant is essential.

- Provide him/her with all relevant details of the business. Remember that they are not piano tuners.
- Record all items of expenditure no matter how small, together with receipts, dates and the reason for them.
- Record all incoming amounts.
- Keep bank statements and cheque stubs.

The Simplex accounting system has proved to be very successful in my own case, together with a ledger for individual details such as invoices, the date they were sent out, and the date paid.

Cash expenditure on petrol/diesel must be accompanied by receipts. This applies even down to small items such as glue or sandpaper, which if not watched can mount up to quite a large sum over the year.

My accountant has complimented me on the simplicity and

logical progression of my accounts but if an accountant expresses a wish for a particular set of books to their own design, please be guided by them. Remember that they are the one who has to argue your point with the tax inspector.

Income tax
Remember: it is up to you to prove your expenditure, not up to the inspector to tell you what your allowances are. This is the province of your accountant. A good accountant is worth his weight in gold, saving far more than his fees over the years.

If good book-keeping is employed, and you are prepared to justify your expenditure with receipts, dates and reasons, the inspector has no room to apply his rules in any other way than the justifiable ones.

Holidays
The employed person's 'perk' of paid holidays is an item which has to be considered by the self-employed and must be covered by the tuning charge or equivalent in repair charges.

If two weeks' summer holidays, bank holidays, and the period in between Christmas and New Year when the piano is covered in mince pies and Christmas cards, are to be considered as average, the charge set against the work conducted during the rest of the year must be at least eight and a half percent.

There is always a considerable amount of Saturday work available throughout the year, from school teachers, business couples and those activities of an entertainment nature such as theatres and concerts. In general, the major music houses and the 10-to-4 tuners do not cover these periods. Time spent in activities such as this can cover the cost of holidays for which no payment is received and for which amounts would have to be put to one side from the normal week's work.

To take holidays during the time of least call for service is also a consideration. The six weeks of the school summer holidays is one of these periods and although the holiday charges are also at their highest and the beaches the most crowded, it is unwise to take them at a peak business time as this is reflected in the six-month reciprocal period when there will be the associated shortage of work. It might be ideal if skiing is your choice and you can go in January/February, the six-month recripocal of the children's school holidays.

By the time all of the above have been taken into account, it may seem that there will be precious little left out of the tuning charge for a take-home-pay, and if tuning charges do not reflect this, you might as well try another job. Take heed and charge the price which will allow the business to provide an income and to prosper.

Peaks and troughs

There have always been peaks and troughs in our business; the secret is to spread the peaks and fill the troughs. Starting in September, when the children go back to school and are settling down to music lessons again, there is a steady build-up of work which lasts until Christmas. If advertising is correct, this build-up will consist of the normal contract work plus new work which has been left until this time.

Customers' promises to children that 'After the holidays, we'll look for a piano', bring in a flood of calls to examine instruments on behalf of prospective purchasers, seeking advice on quality, suitability for a given purpose of pianos on offer. The search goes on until Christmas.

It is often possible to fit these calls in during a working day but sometimes instruments are not available for inspection until the evening. Charges must be made for such calls, depending on the time and travelling involved. They are extra income, and if they are used to fill a vacant period during the day, are extremely useful.

If it is possible to prevent the purchase of an obviously unsuitable instrument at an equally unjustifiable price, these calls can be an excellent customer relations builder.

Many purchasers do not seek the professional advice of a qualified tuner but trust to the local music teacher or pianist to pass judgement on an instrument, based on its performance. When the tuner calls and points out the problems which are beyond the scope of their adviser's knowledge, such phrases as 'Mr Smith said it was suitable' are often the reaction.

My reply is always, 'If I were buying a car, I would take a mechanic with me, not another driver. The problems are of a mechanical nature and as such are not within the scope of pianistic knowledge.'

Pianistic knowledge is in general confined to the 'elephant's teeth' area of the piano and the associated response of them to pianistic requirements. Excursions into technical areas outside their immediate sphere can have disastrous results.

The period after Christmas is quiet, when the piano is probably covered with greetings cards. Workshop repairs estimated for in the September–December period are generally reserved for January–February. The reciprocal six month period of September–December is March–June, and if strenuous efforts are made to encourage the twice-yearly tuning and maintenance, these months will be equally busy.

From the third week in July to the end of August, is a notoriously poor time for household tuning, due to school holidays, with the exception of some school teachers who are not available during term time.

Households where both partners work sometimes have to wait until holiday time to get household things done, as do professional users and of course schools whose instruments are readily available during this period. Repair work and reconditioning saved for just such a period are often a life saver.

Remember, though, that every tuner engaged upon school repairs has the same idea and completion dates for hammer recovering, bass strings, over which one has no control, take a dramatic leap at such times. It is unwise to give completion dates without taking this into account. The wise repairer makes every effort to send off hammers and strings before the holidays commence to ensure an early return.

In all of the foregoing, I have thought of the tuner as being male. At my age, and in a profession which most outsiders consider to be dominated by ancient males and the blind, I have not yet got used to the idea that the up and coming ladies entering our profession will become a delightful addition and can only serve to improve our overall image.

Their problems may be different and with these, I can't help, but I welcome them into the profession whole-heartedly, hoping that my consideration of the general problems common to us all, will meet with their approval.

BIBLIOGRAPHY

Whilst not producing a bibliography for further reading, it might be prudent to mention publications which I have found to be useful without being too technical.

The Modern Piano by Lawrence M. Nalder (ISBN 1 872150 01 2). Published by Heckscher & Co., 75 Bayham Street, London NW1 0AA.

This is a well written and most readable book which takes a logical and progressive look at the aspects of the modern piano of which the average tuner/technician might be required to have a knowledge.

The Anatomy of the Piano by Herbert Shead (ISBN 1 872150 02 0). Published by Heckscher & Co., 75 Bayham Street, London NW1 0AA.

Piano terminology undergoes changes from time to time and the exact description of piano parts and types is extremely useful in order that description of a part or type is understood by everyone.

Piano Servicing, Tuning & Rebuilding by Arthur A. Reblitz (ISBN 0 911572 12 0). Published by The Vestal Press, Vestal, New York 13850. Available from H.J. Fletcher & Newman Ltd., Unit 10A, Industrial Trading Park, Hawley Rd., Dartford, Kent DA1 1QB.

Most students either see or hear of this publication during training and it is a useful guide to the tuning and repairing of pianos. Although dealing mostly with American pianos, the problems are equally applicable to European models.

Pianoforte Tuners' Association, 10 Reculver Road, Herne Bay, Kent CT6 6LD.

The year book contains a wealth of information and a library of interesting publications for further study. The regular seminars and practical classes, although probably at some distance from the main body of tuners, are extremely useful.

INDEX

ACCOUNTANCY 200
ACTION Blüthner Grand 91
ACTION Renner 100
ACTION Richard Check 103
ADJUSTMENT of Grand Pedal 10, 48
ADVERTISING 199
ALEXANDER HERRMANN Upright 1970 95
ALIGNMENT of Flange Rail (Beam) 49
ALIQUOT Patent Blüthner Grand 1913 90
ALLISON Miniature Upright 68
AMYL Grand 1928 15
AREA of Operation 195

BASS TUNING Problems 126, 154
BEAM Loose 49
BEARING Design Problems 125
BECHSTEIN Upright 1882 118
BECHSTEIN Upright 1900 115
BECHSTEIN Upright 1905 120
BENTLY Upright 1964 103
BLÜTHNER GRAND Aliquot Patent 1913 90
BLÜTHNER GRAND Hopper Action 1913 91
BLÜTHNER GRAND 1934 4
BORD UPRIGHT 1894 81
BRASS Compression Flange 81
'BRASSO' Man (The) 19
BROADWOOD UPRIGHT 1935 Bridge Repair 28
BROKEN LOOPS Grand Butt 47
BUSINESS Setting Up The Chapter 9

CASE HISTORIES Chapter 1
CASTORS 191
CHALLEN UPRIGHT 1936 17
CHALLEN UPRIGHT 1976 97
CHAPPELL UPRIGHT 1900/1910 3
CHARGES Tuning 193
CHATTER Cause and Effect 24
COATED STEEL STRINGS 35
COMPRESSION FLANGE Brass 81
COST OF OPERATION 196
CUSTOMER RELATIONS 178

DAEWOO ROYALLE GRAND 1988 100
DAEWOO UPRIGHT 1990 108
DAMP Signs of 6
DAGMAR UPRIGHT 1978 14
DESIGN PROBLEMS Within the Bearing 125
DIAGNOSTIC APPROACH The (Foreword) xi
'DOG-LEG' Key Repair 32
'D'-TYPE Spring & Loop Grand Action Chapter 3
'D' TYPE Spring & Loop Grand Variation 51

EAVESTAFF MINIPIANO 1936 55
EAVESTAFF MINIPIANO 1937 61
EAVESTAFF MINIPIANO ROYAL 1939 63
EAVESTAFF MINIPIANO ROYAL 1957 66
ELECTRONIC TUNING AIDS 139
EQUAL TEMPERAMENT 144
ERROR Searching the Bearing for 149

F. MENZEL UPRIGHT 1890 73
FLANGE Brass Compression 81
FLANGE RAIL Loose 49
FLANGE Underdamper 31

GAINING ACCESS Allison Miniature 1939 69
GAINING ACCESS Eavestaff Minipiano 1937 61
GAINING ACCESS Eavestaff Minipiano Royal 1939 64
GAINING ACCESS Eavestaff Minipiano Royal 1957 66
GAINING ACCESS Kemble Minx 1959 70
GAINING ACCESS Marshall & Rose Grand 1936 78
GLUE Sticking Hammers 4
GLUE Over Enthusiastic Use of 4
GRAND ACTION Spring & Loop 'D' Type Chapter 3
GRAND AMYL 1928 15
GRAND BLÜTHNER Aliquot Patent 1913 90
GRAND BLÜTHNER 1934 4
GRAND DAEWOO 1988 100
GRAND HAAKE 1890 87
GRAND LEG REPAIR 34
GRAND MARSHALL & ROSE 1936 78
GRAND PEDAL ADJUSTMENT 10, 48
GRAND PLEYEL WOLF LYON 1902 75
GRAND STEINWAY 1921 New York 10
GRAND STEINWAY Model B 1971 116
GRAND TROLLEY FRAMES 191
GUSTAF WEISCHEL UPRIGHT 84

HAAKE GRAND 1890 87
HAMMER HEADS Loose 18
HAMMER REFACING/RESHAPING 39
HAMMER RESPACING 42
HAMMER SHANKS Red Cedar Wood 3
HAMMER TRAVEL 42
HEARING FOCUS 153
HERRMANN Alexander Upright 95
HISTORIES Case Chapter 1
HOLIDAYS 201
HOPPER ACTION Blüthner Grand 91
HOPPER PATENT 114 1896 81
HUMIDITY Chapter 7

INCOME TAX 201
INSTRUMENTS Unusual and Specialist Chapter 5
INSURANCE Loss of Earnings 197
INTERVAL The 145
INTONATIONS Pythagorean 155, 157
IRREGULAR MEANTONE TEMPERAMENT 170

JACK Tail-less Upright 80
JEREMY Ode To 181

KEMBLE Minx Miniature Upright 1959 70
KEMBLE UPRIGHT 1988 102
KEYBOARD TEMPERAMENTS 144
KEY COVERING Synthetic 36
KEY REPAIR Broken 32
KEY SLIP Grand 9

KEY SLIP Upright 9

LEARNING TO SEE (Foreword) xi
LINDNER UPRIGHT 1970 77
LOCK RAIL Upright 9
LOOPS Repair 47
LOOSE Hammer Heads 18
LOSS OF EARNINGS INSURANCE 197

MARSHALL & ROSE GRAND 1934 78
MATERIAL Poor Quality 45
MEANTONE TEMPERAMENTS 161
MENZEL F. Upright 1890 73
MICE Signs of 7
MINIATURE PIANOS Chapter 4
MINIATURE UPRIGHT Allison 1939 68
MINIATURE UPRIGHT Kemble Minx 1959 70
MINIPIANO Eavestaff 1936 55
MINIPIANO Eavestaff 1937 61
MINIPIANO Eavestaff ROYAL 1939 63
MINIPIANO Eavestaff 1957 66
MONINGTON & WESTON UPRIGHT 1930 115

OCTAVE BALANCE The Laws of 150
OCTAVE STRETCH 153
ODE TO JEREMY 181
OPERATION Area of 193
OPERATION Cost of 196
ORDINAIRE Temperament 172

PEAKS & TROUGHS 202
PEDAL ADJUSTMENT Grand 10, 48
PENSION Retirement 198
PLAYING TO TUNING RATIO 142
PLEYEL WOLF LYON GRAND 1902 75
POOR QUALITY MATERIALS 45
PRESENTATION 175
PRESSURE BAR Steinway 11
PUBLIC LIABILITY INSURANCE 198
PYTHAGOREAN INTONATIONS 155, 157

RAIL Loose Flange 49
RAIL LOCK Upright 9
RANGE OF TOOLS CARRIED 190
RED CEDAR WOOD HAMMER SHANKS 3
REFACING/RESHAPING OF HAMMERS 39
REGULATING HOPPER Patent 114 1869 81
RELATIONS Customer 178
RENNER ACTION Daewoo Royalle Grand 100
REPAIRING BROKEN LOOPS Hammer Butt 47
REPAIRING GRAND PIANO LEG 34
REPLACEMENT OF UNDERDAMPER FLANGES 31
RESPACING OF HAMMERS 42
RETIREMENT PENSION 198
RICHARD CHECK ACTION 103
ROYAL Eavestaff Minipiano 1939 63
ROYAL Eavestaff Minipiano 1957 66
ROYALLE GRAND Daewoo 1988 100

SAMES UPRIGHT 1910 113
SEARCHING THE BEARING FOR ERROR 149
SECRETARIAL TIME 199
SELF INFLICTED TUNING PROBLEMS 123
SETTING UP THE BUSINESS Chapter 9
SIGNS OF MICE 7
SMALL TUNING HEAD Use of 138
SPECIALIST & UNUSUAL INSTRUMENTS Chapter 5
SPEED TEST 153

SPLIT BRIDGE REPAIR 28
SPLIT WREST PLANK 1
SPRING & LOOP GRAND ACTION 'D' Type Chapter 3
STARR UPRIGHT 1987 110
STEEL STRINGS Coated 35
STEINMAN UPRIGHT 1987 106
STEINWAY GRAND 1921 New York 10
STEINWAY UPRIGHT 1894 11
STEINWAY UPRIGHT 1937 13
STICKING HAMMERS 7
SYMPATHETIC VIBRATION 133
SYNTHETIC KEY COVERINGS 36

TAIL-LESS UPRIGHT JACK 50
TELEPHONE 198
TEMPERAMENT Equal 144
TEMPERAMENT ORDINAIRE 172
TEMPERAMENTS KEYBOARD 144
TEMPERAMENTS MEANTONE 161
TOOLS Range Carried 190
TRAVEL Hammer 42
TUNING AIDS 139
TUNING CHARGES 193
TUNING HEAD Small 138
TUNING PROBLEMS Bass 126
TUNING PROBLEMS Treble 128
TUNING WEDGE 129

UNDERDAMPER FLANGES Replacement 31
UNUSUAL & SPECIALIST INSTRUMENTS Chapter 5
UPRIGHT Allison Miniature 1939 68
UPRIGHT Alexander Herrmann 1970 95
UPRIGHT Bechstein 1882 118
UPRIGHT Bechstein 1900 115
UPRIGHT Bechstein 1905 120
UPRIGHT BENTLY 1964 103
UPRIGHT BORD 1894 81
UPRIGHT BROADWOOD 1935 28
UPRIGHT CHALLEN 1936 17
UPRIGHT CHALLEN 1976 97
UPRIGHT CHAPPELL 1900/1910 3
UPRIGHT DAEWOO 1900 108
UPRIGHT DAGMAR 1978 14
UPRIGHT GUSTAF WEISCHEL 84
UPRIGHT KEMBLE MINX MINIATURE 1959 70
UPRIGHT KEMBLE 1988 102
UPRIGHT LINDNER 1970 77
UPRIGHT MENZEL F. 1890 73
UPRIGHT MONINGTON & WESTON 1930 115
UPRIGHT SAMES 1910 113
UPRIGHT STEINWAY 1894 11
UPRIGHT STEINWAY 1937 13
UPRIGHT STARR 1987 110
UPRIGHT TAIL-LESS JACK 30
UPRIGHT WADDINGTON 1920 114
UPRIGHT WILHELM STEINMAN 1987 106
UPRIGHT YAMAHA Model 1986 117

VARIATION On 'D' Type Action 51
VIBRATION Sympathetic 133

WARPED Lock Rail Upright 46
WATER DAMAGE 31
WARPED KEY SLIP Grand 46
WADDINGTON UPRIGHT 1920 114
WEDGE Tuning 129

YAMAHA UPRIGHT Model U1A 1986 117